T0212874

SpringerBriefs in Applied Sciences and Technology

More information about this series at http://www.springer.com/series/8884

Wolfgang Gräfe

Quantum Mechanical Models of Metal Surfaces and Nanoparticles

 Springer

Wolfgang Gräfe
Berlin
Germany

ISSN 2191-530X   ISSN 2191-5318   (electronic)
SpringerBriefs in Applied Sciences and Technology
ISBN 978-3-319-19763-0   ISBN 978-3-319-19764-7   (eBook)
DOI 10.1007/978-3-319-19764-7

Library of Congress Control Number: 2015941358

Springer Cham Heidelberg New York Dordrecht London

Printed on acid-free paper

Springer International Publishing AG Switzerland is part of Springer Science+Business Media
(www.springer.com)

Preface

In this book I consider two simple quantum mechanical models of metal surfaces. It is the aim to give an ostensive picture of the forces acting in a metal surface and to deduce analytical formulae for the description of their physical properties. The starting points of my approach to the surface physics were strength and fatigue limit. As the cause of these features I consider a near-surface stress with the dimension of a force per area. In this book I explain the relation between the near-surface stress and the familiar surface parameters.

In order to make the understanding of my theory easier I have applied the concept of the separation of the three-dimensional body into three one-dimensional subsystems.

This book has been written for experts and newcomers in the field of surface physics.

Wolfgang Gräfe

Acknowledgments

Without the patience and without the care of my wife Herta I would not have accomplished this book.

Contents

Nomenclature

a	Width of potential well
b	Width of potential barrier
c	Lattice constant
A	Surface area
d_e	Electron density
e	Absolute value of electron charge
E	Energy
E^B	Bottom of the energy band
E^F	Fermi-level
E^S	Energy of the surface state
E^{Sa}	Energy of an additional surface state
E^{So}	Energy of an offspring surface state
E^T	Top of the energy band
E^{ul}	Energy level in an unlimited (infinitely extended) body
ESB	Electrons in the surface band (superscript)
f_i	ith Component of the force per electron
F	Force
F_i^S	ith Component of the force in the surface layer
$\hbar$	Reduced Planck's constant
i	Imaginary unit
i	Index
j	Index
k	Index
k	Boltzmann's constant
k	Wave number
L	Length of an edge
m	Electron mass
n	Density of electrons
N	Number of electrons
N^A	Number of an atom
N^{el}	Total number of energy levels

p	Fermi distribution function, probability
pl	Plate-like (superscript)
q	Surface charge density
q^{ESB}	Surface charge resulting from the electrons in the surface band
r	Radius
R	Radius
s_{ij}	Near-surface stress
S	Surface state (superscript)
Sa	Additional surface state (superscript)
So	Offspring surface state (superscript)
T	Absolute temperature
TR	Transition (superscript)
ul	Unlimited (superscript)
U	Potential barrier in the bulk
U^S	Potential barrier at the surface
x	Coordinate
X	Number of mols
Y	Young's modulus, modulus of elasticity
Z	Number of particles
Z	Partition function
γ	Surface tension
δ	Attenuation length, penetration depth of a wave function
Δ	Laplace operator
$\delta(x)$	Delta function
ε	Dielectric coefficient
ε	Strain
ε_{ij}	Strain tensor
ε_0	Absolute permittivity
ϕ	Potential difference across the interface
φ	Surface free energy
φ^{ESB}	Surface energy resulting from the electrons in the surface band
φ^{TR}	Contribution to the surface free energy due to electron transition
μ	Index
ν	Index
ψ	Wave function
ς_{ij}	Estance, surface stress-charge coefficient
σ_{ij}	Surface stress tensor
σ_{ij}^{ESB}	Surface stress tensor resulting from the electrons in the surface band
ζ	Chemical potential
Ω	Volume of the phase space

Chapter 1
Introduction

Abstract The surface free energy is the work performed from outside for the generation of an additional surface. The surface tension is the force per length acting on the surface. For a liquid the amount of the surface free energy per unit area equals the surface tension. Electrocapillarity means the change of the surface tension due to the influence of a surface charge. The surface tension reaches a maximum if the surface charge density vanishes. For a solid the quantities "surface free energy" and "surface tension" are different. Shuttleworth has formulated a relation between the surface free energy per unit area φ and the surface stress σ

$$\sigma = \varphi + A \left(\frac{d\varphi}{dA} \right)_T.$$

A compilation of experimental and theoretical data in the literature is given for the surface parameters of solids. It is the aim of the book to classify the notion "near-surface stress" in the list of surface quantities. The near-surface stress has been used by the author for an explanation of the cause of fatigue limit in strength investigations.

Keywords Surface energy of solids · Surface stress of solids

The surface free energy is the work performed from outside for the generation of an additional surface. The surface tension is the force per length acting on the surface. For a liquid the amount of the surface free energy per unit area equals the surface tension. As a matter of principle, for a liquid the experimental determination of the surface tension is a straightforward procedure.

1.1 Electrocapillarity of Liquids

In the case of liquids, electrocapillarity means the change of the surface tension due to the influence of a surface charge.

© The Author(s) 2015
W. Gräfe, *Quantum Mechanical Models of Metal Surfaces and Nanoparticles*,
SpringerBriefs in Applied Sciences and Technology,
DOI 10.1007/978-3-319-19764-7_1

If the difference of the electric potential across the interface between a mercury electrode and a surrounding electrolyte is varied, the surface tension γ of the mercury changes too. For a certain potential difference ϕ across the interface, the surface charge density q in the mercury is zero. According to the Lippmann–Helmholtz equation

$$\left(\frac{\partial \gamma}{\partial \phi}\right)_{p,T,\zeta} = -q^M = q^L \tag{1.1}$$

the surface charge density is referred to the surface tension γ. The surface tension reaches a maximum if the surface charge density vanishes. (For the Lippmann–Helmholtz equation, see Kortüm [1].) The symbol ζ means the chemical potential, q^M and q^L are the excess charge densities in the metal and in the electrolyte at the phase interface. In the experimental situation considered here, the surface tension γ is the Gibbs surface free energy per surface area.

Due to the mutual repulsion of the charges in the surface layer a negative contribution to the surface energy and the surface stress will arise which is zero for a vanishing surface charge density q.

1.2 Surface Free Energy and Surface Stress of Solids

As Gibbs [2] has pointed out, for a solid the quantities "surface free energy" and "surface tension" are different at least in their nature. Here too, the surface free energy is the work performed from outside in the production of an additional surface. Furthermore, we have to consider stresses and strains in the solid surface by which no additional atom will be introduced into the surface. It is always possible to find two perpendicular directions in the surface for which no shear stresses exist. The surface stress components in these two directions are the principal surface stresses. For the case of a solid, Shuttleworth [3] has defined the surface tension as the arithmetic mean of the values of the principal surface stresses. According to him (citation), "for an isotropic substance, or for a crystal face with a three- (or greater-) fold axis of symmetry, all normal components of the surface stress equal the surface tension."

Shuttleworth [3] has formulated the following thermodynamic relation between the quantities, surface free energy per unit area φ and surface tension σ.

$$\sigma = \varphi + A\left(\frac{\mathrm{d}\varphi}{\mathrm{d}A}\right)_T. \tag{1.2}$$

According to this relation, the surface tension σ is the sum of the surface free energy per unit area φ and its strain derivative. The letter A means the surface area.

For the surface stress tensor σ_{ij} with the dimension of a force per length, Herring [4] deduced the formula

$$\sigma_{ij} = \varphi \delta_{ij} + \frac{\partial \varphi}{\partial \varepsilon_{ij}}. \tag{1.3}$$

Here δ_{ij} is the Kronecker symbol and ε_{ij} is the surface elastic strain tensor. In both second rank tensors ε_{ij} and σ_{ij} the indices i and j take only two values, e.g., 1 and 2.

For the symmetries and restrictions discussed above, the Eq. (1.3) becomes identical to Eq. (1.2).

Depending on the experimental conditions, the notion "surface free energy" can mean the Gibbs surface free energy or the Helmholtz surface free energy. For a more detailed discussion of the thermodynamic potential needed, see Ibach [5].

Considering a solid, we continue to use instead of the notion "surface tension" only the notions "surface free energy per unit area φ" and "surface stress σ" as has been recommended by Cammarata [6] and also has been exercised in essence by Haiss [7] in the their review articles.

Remark

In *mechanics* a shortage of the bonds between the atoms is caused by compressive stresses. The compressive stresses are characterized by a negative sign.

In *surface science* a shortage of the bonds between the atoms in the surface layers is caused by a tensile surface stress. This tensile surface stress is denoted by a positive sign as in mechanics. (Tensile surface stresses are related to higher strengths of the solids.)

1.3 The Estance or the Surface Stress-Charge Coefficient

Gochstein [8] introduced the quantity "estance". This notion is the abbreviation of the words "elastic" and "impedance". The estance is a measure of the change in the surface stress caused by a variation of a surface charge in a solid which is in contact with an electrolyte. He has defined two variants, a q-estance meaning $\partial \sigma / \partial q$ and the ϕ-estance which stands for $\partial \sigma / \partial \phi$. Here too, ϕ is the potential difference across the boundary between the solid and the electrolyte. (Gochstein uses the symbol γ instead of σ but, he has in mind the surface stress.) In the anglophone literature the differential quotient $\partial \sigma / \partial q$ has the designation "surface stress-charge coefficient".

According to Kramer and Weissmüller [9] the Lippmann–Helmholtz equation, Eq. (1.1), "applies equally to solids and fluids" and "is an excellent approximation in both cases, except when the specific surface area is extremely large."

1.4 Experimental Data in the Literature

According to Landolt and Boernstein [10] the surface free energy of sodium at its melting point is 0.44 J/m^2.

For the experimental determination of the surface stress on some metals Vermaak and coworkers [11–13] measured the radial strain in small spheres by electron diffraction and calculated the average surface stress. For Cu, Ag, Au, and Pt they have obtained values ranging from 1.175 N/m for Au via 1.415 N/m for Ag to 2.574 N/m for Pt. The surface stress for Cu has a value between ±0.45 N/m.

From anomalies in the dispersion of surface phonons in a clean Ni (110) surface Lehwald et al. [14] have determined surface stresses of 4.2 and 2.1 N/m for two different crystallographic directions.

Haiss et al. [15] interpret their results as revealing a strong linearity between stress and charge. For an Au(111) surface in contact with different electrolytes they have found values for the surface stress-charge coefficient (estance) $\varsigma_{ij} = \partial \sigma_{ij}/\partial q$ ranging from −0.67 to −0.91 V.

Ibach [16] determined for Au(111) surfaces a linear relationship between stress and charge with $\varsigma = -0.83$ V. In contrast to this, for an Au(100) surface the stress-versus-charge curve could only be fitted with a parabola.

Weissmüller and coworkers [17] prepared nanoporous Pt samples and brought them into contact with electrolytes. They varied the interface charge by applying an electric field across the interface. They measured in situ the variation of strain in the samples by a dilatometer and by X-ray diffraction. In this way, the authors have determined the values of −0.7 and −1.6 V for the surface stress-charge coefficient (estance). Also Weissmüller et al. [18] observed a macroscopic contraction of nanoporous Au caused by the influence of a negative surface charge. For gold, the authors have realized negative values of ς_{ij}.

Viswanath et al. [19] have reported experimental studies on nanoporous Pt immersed in aqueous solutions of NaF. According to their results, the surface stress-charge coefficient (estance) varies for concentrations X with 0.02 M < X < 1 M from −1.9 V to less than the half of this value at $X = 1$ M.

The charge density variation in experiments performed by Haiss [7] revealing a strong linearity between stress and charge density is 2×10^{-1} C/m^2. Elsewhere, Haiss et al. [15] have reported a charge density variation smaller than 2.5×10^{-1} C/m^2. (In fact, Weissmüller and coworkers [17] refer a variation of the overall charge density of 5 C/m^2, however for a nanoporous material!)

1.5 State of the Theoretical Knowledge

In the literature, there are a lot of semiempirical as well as first principles calculations for the determination of the surface stresses for nonconductors, for the semiconductors Ge, Si, and A^{III}B^V compounds, and for some metals.

On a semiempirical base, Tyson and Miller [20] have derived a relation between the specific surface energy of a solid metal in contact with its vapor φ^{SV} and the liquid–vapor surface energy of the same metal φ^{LV}, both at the melting point T^m. The authors also have estimated the surface entropy and have given a formula for the calculation of φ^{SV} at any temperature between 0 K and T^m. The authors have listed values of φ^{SV} at the absolute zero of temperature and at the melting point for a series of metals. The highest surface energy amounting to 3.25 J/m^2 has been determined for tungsten at 0 K.

Miedema [21] has developed a model in which the atomic bond energy in solids is interpreted as the surface energy of the atoms. He has determined the surface energies for metals at the temperature 0 K from experimental values of the surface energies at the melting points. The highest value he reported is found for Rhenium and amounts to 3.65 J/m^2. For Na the surface energy at the absolute zero is 0.26 J/m^2.

Wolf and Griffith [22] discussed the physical origin of the difference between surface free energy and surface tension at the surface of a crystal in terms of a simple model of rigid parallel planes with phenomenological bulk and surface energies. In their derivations also a near-surface local stress appeared.

Using the simple empirical n-body Finnis–Sinclair potentials, Ackland et al. [23, 24] have calculated the surface free energy and the surface stress for fcc and bcc metals. The values of the surface free energy range from 0.62 J/m^2 for an Ag (111) surface to 3.036 J/m^2 for a (310) surface on W. For the principal surface stresses the authors have obtained values between 0.263 N/m for V and 3.085 N/m for Ta.

Joubert [25] showed that the high density of electronic surface states near the Fermi level gives rise to an enhanced attractive interaction between neighboring pairs of atoms on the (001)-surface of tungsten.

Needs [26] has calculated the tensor of surface stress at aluminum surfaces. He performed self-consistent local-density-functional calculations using norm-conserving pseudopotentials. According to his assumptions the surface layer of a crystal may reduce its energy by relaxation of the atomic layers. The lowest energy configuration of the crystal will have the surface layer stressed in its own plane, whereas the bulk of the material exerts an opposing stress so that equilibrium is maintained. The calculated surface stresses are tensile and range from +0.145 eV/Å^2 (+2.32 N/m) for the (111) surface to +0.124 eV/Å^2 (+1.99 N/m) and +0.115 eV/Å^2 (+1.84 N/m) for the (110) surface. That means the surface favors contraction in its plane and as a result the bulk is under compression.

Needs and coworkers [27, 28] also calculated the surface free energy and the surface stress for clean and unreconstructed (111) surfaces of the fcc metals Al, Au, Ir, and Pt and have obtained for the surface energy values between 0.96 and 3.26 J/m^2. The surface stresses range from 0.82 to 5.60 N/m.

Gräfe [29] considered a "near-surface stress". That is a stress with the dimension of a force per area which is concentrated in a layer near the surface. Its magnitude decays in the direction normal to the surface. The near-surface stress is inherently related to the surface stress.

Wolf [30] applied the many-body potential of the embedded-atom method and the Lennard-Jones potential to the calculations of the surface energies for 85 different surfaces on fcc and bcc metals.

Feibelman [31] used a parallel, linear combination of atomic orbitals (LCAO) implementation of the local-density approximation (LDA). The Pt (111) surface has been modeled by a 9-layer (111) slab. The two outer atomic layers on either side of the slab were allowed to relax. For a clean (111) surface of Pt he has determined a tensile surface stress constituting 392 meV/Å^2 (6.297 N/m). Feibelman has declared that there are no rigorous theorems concerning a systematic of surface stresses. But, the calculations for clean metal surfaces carried out so far, result in tensile stresses.

Friesen et al. [32] have reported the results of calculations with first principles methods for the determination of surface stresses. The values amount to 2.77 N/m for Au(111) and 0.82 N/m for Pb(111).

Umeno et al. [33] determined the scalar surface stress-charge coefficient (estance) $\varsigma_{ij} = \partial\sigma_{ij}/\partial q$ (the trace of the tensor) for gold by an analysis of the strain dependence of the work function. These calculations were realized by applying the density functional theory. For the (111), (110) and (100) surfaces of Au, they have obtained values between -1.86 and 0 V.

1.6 The Aim of the Following Text

It is the aim of the following text to classify of the notion "near-surface stress" in the list of physical quantities describing the surface physical phenomena. This quantity has been used by Gräfe [29] for an explanation of the cause of fatigue limit in strength investigations. Therefore, in this booklet the main concern is with metal surfaces. For a more detailed discussion of the relation between fatigue limit and near-surface stress see Eqs. (4.6) and (4.7) as well as Sect. 14.4.

References

1. Kortüm G (1972) Lehrbuch der Elektrochemie, Verlag Chemie Weinheim, p 397
2. Gibbs JW (1961) The scientific papers 1, thermodynamics. Dover Publications New York, p 315 (Longmans, Green and Co, London 1906)
3. Shuttleworth R (1950) The surface tension of solids. Proc Phys Soc (Lond) A63:444–457
4. Herring C (1951) Surface tension as a motivation for sintering. In: Kingston WE (ed) The physics of powder metallurgy. McGraw-Hill, New York, pp 165, 143–180
5. Ibach H (2006) Physics of surfaces and interfaces. Springer, Berlin, p 161
6. Cammarata RC (1994) Surface and interface stress effects in thin films. Prog Surf Sci 46:1–38
7. Haiss W (2001) Surface stress on clean and adsorbate-covered solids. Rep Prog Phys 64:591–648
8. Gochshtejn A (1976) Poverchnostnoe natjazhenie tverdych tel i adsorbcija, Izd. Nauka Moskva, p 15, Chap 4
9. Kramer D, Weissmüller J (2007) A note on surface stress and surface tension and their interrelation via Shuttleworth's equation and the Lippmann equation. Surf Sci 601:3042–3051
10. K.Schäfer (ed) (1968) Landoldt-Börnstein Bd. II/5b, Eigenschaften der Materie in ihren Aggregatzuständen, 5. Teil, Bandteil b, Transportphänomene II—Kinetik; Homogene Gasgleichgewichte, 6st edn. Springer, Berlin, pp 9–11
11. Mays CW, Vermaak JS, Kuhlmann-Wilsdorf D (1968) On surface stress and surface tension: II. Determination of the surface stress of Gold. Surf Sci 12:134–140

12. Wassermann HJ, Vermaak JS (1970) On the determination of lattice contraction in very small silver particles. Surf Sci 22:164–172
13. Wassermann HJ, Vermaak JS (1972) On the determination of the surface stress of copper and platinium. Surf Sci 32:168–174
14. Lehwald S, Wolf F, Ibach H, Hall BM, Mills DL (1987) Surface vibrations on Ni(110): the role of surface stress. Surf Sci 192:131–162
15. Haiss W, Nichols RJ, Sass JK, Charle KP (1998) Linear correlation between surface stress and surface charge in anion adsorption on Au(111). J Electroanal Chem 452:199–202
16. Ibach H (1999) Stress in densely packed adsorbate layers and stress at the solid-liquid interface —Is the stress due to repulsive interactions between the adsorbed species? Electrochim Acta 45:575–581
17. Weissmüller J, Viswanath RN, Kramer D, Zimmer P, Würschum R, Gleiter H (2003) Charge-induced reversible strain. Science 300:312–315
18. Kramer D, Viswanath RN, Weissmüller J (2004) Surface-stress induced macroscopic bending of nanoporous gold cantilevers. Nano Lett 4:793–796
19. Viswanath RN, Kramer D, Weissmüller J (2005) Variation of the surface stress-charge coefficient of platinum with electrolyte concentration. Langmuir 21:4604–4609
20. Tyson WR, Miller WA (1977) Surface free energies of solid metals: estimation from liquid surface tension. Surf Sci 62:267–276
21. Miedema AR (1979) Das Atom als Baustein in der Metallkunde. Philips Technol Rundsch 38:269–281
22. Wolf DE, Griffith RB (1985) Surface tension and stress in solids: the Rigid-Planes model. Phys Rev B 32:3194–3202
23. Ackland GJ, Finnis MW (1986) Semi-empirical calculation of solid surface tensions in body-centred cubic transition metals. Philos Mag A 54:301–315
24. Ackland GJ, Tichy G, Vitek V, Finnis MW (1987) Simple N-body potentials for the noble metals and nickel. Philos Mag A 56:735–756
25. Joubert DP (1987) Electronic structure and the attractive interaction between atoms on the (001) surface of W. J Phys C: Solid State Phys 20:1899–1907
26. Needs RJ (1987) Calculations of the surface stress tensor at aluminum (111) and (110) surfaces. Phys Rev Lett 58:53–56
27. Needs RJ, Godfrey MJ (1990) Surface stress of aluminum and jellium. Phys Rev B 42:10933–10939
28. Needs RJ, Godfrey MJ, Mansfield M (1991) Theory of surface stress and surface reconstruction. Surf Sci 242:215–221
29. Gräfe W (1989) A surface-near stress resulting from Tamm's surface states. Cryst Res Technol 24:879–886
30. Wolf D (1990) Correlation between energy, surface tension and structure of free surfaces in fcc metals. Surf Sci. 226:389–406
31. Feibelman PJ (1997) First-principles calculations of stress induced by gas adsorption on Pt (111). Phys Rev B 56 (1997-II):2175–2182
32. Friesen C, Dimitrov N, Cammarata RC, Siradzki K (2001) Surface stress and electrocapillarity of solid electrodes. Langmuir 17:807–815
33. Umeno Y, Elsässer C, Meyer B, Gumbsch P, Nothacker M, Weissmüller J, Evers F (2007) Ab initio study of surface stress response to charging, EPL 78:13001-p1–13001-p-5

Chapter 2
The Model of Kronig and Penney

Abstract Kronig and Penney considered a Meander-like potential energy of the electrons $U(x)$ extended from $-\infty$ until $+\infty$ as a one-dimensional model of a solid. For this model the probability density of electron states has been calculated.

Keyword Kronig-Penney model

As a one-dimensional model of a solid, Kronig and Penney [1] considered a Meander-like potential energy of the electrons $U(x)$ extended from $-\infty$ until $+\infty$. The mathematical description of the potential energy is

$$U(x') = 0 \quad \text{for} \quad 0 < x' < a \tag{2.1}$$

and

$$U(x') = U \quad \text{for} \quad a \leq x' \leq (a + b) = c \tag{2.2}$$

with $x' = x - c\,(N^A - 1)$. The quantity N^A means the number of an atom arranged in the x-direction.

The one-dimensional, time-independent Schrödinger equation for the wave function ψ of the electrons in the considered potential is

$$-\frac{\hbar^2}{2m}\Delta\psi + U(x)\psi = E\psi. \tag{2.3}$$

The symbol $\hbar$ means the reduced Planck's constant, m the mass of an electron, and Δ the Laplace operator, respectively.

The allowed energy levels E of the electrons in the bulk of a body with a periodical potential are placed in the allowed energy bands.

W. Gräfe, *Quantum Mechanical Models of Metal Surfaces and Nanoparticles*,
SpringerBriefs in Applied Sciences and Technology,
DOI 10.1007/978-3-319-19764-7_2

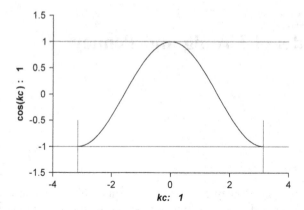

Fig. 2.1 The *thick line* shows the run of cos (*kc*). The *horizontal lines* at the ordinates +1 and −1 are the values of *g*(*E*, *U*, *a*, *b*) belonging to the lower and the upper boundaries of the allowed energy band. The *short vertical lines* mark the values *kc* = ± π

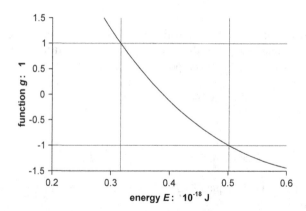

Fig. 2.2 The *thick line* depicts the run of the function *g*(*E*, *U*, *a*, *b*). The *horizontal lines* at +1 and −1 are the values of *g*(*E*, *U*, *a*, *b*) at the boundaries of the allowed energy band. The *vertical lines* mark the lower and the upper boundaries of the allowed energy band

In the energy range $U > E$ follows from the matching conditions for the wave functions the Eq. (2.4)

$$\cos kc = \cos \beta a \times \cosh \gamma b - \frac{\beta^2 - \gamma^2}{2\beta\gamma} \sin \beta a \times \sinh \gamma b = g(E, U, a, b). \qquad (2.4)$$

Here k means the wave number, $\beta^2 = \kappa^2 E$, and $\gamma^2 = \kappa^2(U - E)$, respectively. The symbol κ^2 stands for $2\,m/\hbar^2$.

The fulfillment of Eq. (2.4) is visualized in Figs. 2.1 and 2.2.

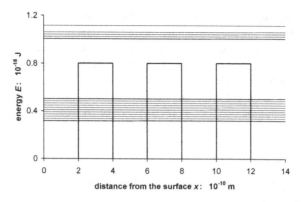

Fig. 2.3 Schematic representation of the run of potential energy in the bulk and of the energy levels in the allowed energy bands (*shaded stripes*) calculated for a one-dimensional body; (Here, the *thin lines* mark only the range of the allowed energy bands and have no physical meaning.)

In order to find the values of $\pm kc$ corresponding to the energy levels in the allowed energy band, we start in Fig. 2.2 from the abscissa and move vertically upward. From the cross point with the thick curve we shift horizontally to the thick line in Fig. 2.1 and from there vertically down to the abscissa.

In Fig. 2.1, the values of kc corresponding to the energy levels in the allowed energy band are located within the interval

$$-\pi \leq kc \leq \pi. \tag{2.5}$$

The energy eigenvalues have been calculated for $a = 2 \times 10^{-10}$ m, $b = 2 \times 10^{-10}$ m, and $U = 0.8 \times 10^{-18}$ J (5 eV). The energies of the edges of the energy bands are for

- the bottom of the lower energy band: $E^{Bl} = 0.317 \times 10^{-18}$ J.
- the top of the lower energy band: $E^{Tl} = 0.502 \times 10^{-18}$ J.
- the bottom of the upper energy band: $E^{Bu} = 1.005 \times 10^{-18}$ J.
- the top of the upper energy band: $E^{Tu} = 1.880 \times 10^{-18}$ J.

The three lowest band edges are depicted in Fig. 2.3.

2.1 The Density of the Electron Energy Levels $n(E)$

The motion of the quasi-free electrons in a periodic potential is described by the wave number k, defined by

$$k = \frac{2\pi}{L} n. \tag{2.6}$$

Generally speaking, L is the observation abstraction for the length of an oscillating system. In the infinitely extended body, the quantity L means an arbitrary length the one-dimensional body is subdivided in. With other words, the infinitely extended body is composed by a periodical repetition of subranges with the length L in the x-direction. The value of L limits the number of the wave functions per energy band but it does not limit the extension of the wave functions.

For each wave number vector an oppositely directed wave number vector exists

$$\left|\vec{k}\right| = \left|\vec{\bar{k}}\right|. \tag{2.7}$$

Without regard to the spin, it follows for the number of the energy levels in the allowed energy band

$$n = 2\frac{L}{2\pi}k. \tag{2.8}$$

By differentiation of Eq. (2.8) with respect to the energy E we obtain for the density of states in an allowed energy band

$$n(E) = \frac{dn}{dE} = \frac{L}{\pi}\frac{dk}{dE}. \tag{2.9}$$

In an infinitely extended solid with its infinite number of atoms, infinitely many different values of the wave number k exist. Therefore, the density of the allowed energy levels is infinitely large too. For this situation, it is only meaningful to use the probability density for the existence of an electron state with the energy E

$$p(E) = \lim_{N\to\infty}\frac{1}{N}\frac{dn}{dE} = \lim_{N\to\infty}\frac{1}{N}\frac{L}{\pi}\frac{dk}{dE}. \tag{2.10}$$

Figure 2.4 shows the probability density of electron states $(c/L)n(E)$ for the energy E.

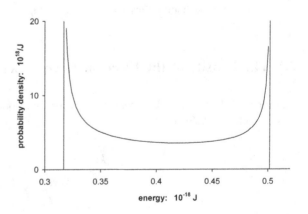

Fig. 2.4 The probability density of electron states $(c/L)n(E)$ for a one-dimensional body versus the energy E

2.2 Remarks

The mean value $(a + b)/2 = 2 \times 10^{-10}$ m is comparable with the radius of sodium atoms. That means, the number of atoms along an edge of a cube with a length of $L = 1$ m is 2.5×10^9 atoms.

For the potential barrier U in the bulk a value near the ionization potential of sodium has been chosen.

The applied method for the theoretical calculations is a separable crystal potential in the one-electron Schrödinger equation.

In the Kronig–Penney model there is no statement concerning the position of the Fermi level. Therefore, the model of the solid can be adapted to bodies with different physical properties.

Reference

1. de Kronig RL, Penney WG (1931) Quantum mechanics of electrons in crystal lattices. Proc R Soc 130(Ser A):499–513

Chapter 3
Tamm's Electronic Surface States

Abstract The one-dimensional body shall have a surface at $x = 0$. This surface is physically described by a step of the potential energy U^S. As Tamm has shown, electrons can be localized at the surface of the body. The energies E and the attenuation lengths δ for the two resulting Tamm's electronic surface states have been calculated.

Keyword Surface states

The one-dimensional body shall have a surface at $x = 0$. This surface is physically described by a step of the potential energy $U(x)$ of the height U^S with $U^S > U$ as it is depicted schematically in Fig. 3.1.

As Tamm [1] has shown, electrons can be localized at the surface of the body.

The energy levels of the electronic surface states are determined by two conditions. The first condition is the decay of the wave functions in the direction into the bulk. This is only possible if the energy levels are located within the forbidden energy bands of the bulk. In these energy gaps it is $g(E, U, a, b) > 0$. In order to fulfill Eq. (2.4), the wave number k must be replaced by $i/\delta + \pi/c$. The quantity δ is a measure of the extension of the wave function into the bulk. With this transformation of k we obtain instead of Eq. (2.4)

$$\operatorname{sign}(g)\cosh\frac{c}{\delta} = \cos\beta a \times \cosh\gamma b - \frac{\beta^2 - \gamma^2}{2\beta\gamma}\sin\beta a \times \sinh\gamma b = g(E, U, a, b).$$

(3.1)

The second condition is the matching of the wave function in the bulk with the wave function beyond $x = 0$. From this condition and for $E < U$ it follows the Eq. (3.2),

$$\frac{1}{\delta} = \frac{1}{c}\ln\left(\operatorname{sign}(g)\frac{\beta\gamma\cosh\gamma b - \sqrt{\kappa^2 U^S - \beta^2}\,\beta\sinh\gamma b}{\sqrt{\kappa^2 U^S - \beta^2}\,\gamma\sin\beta a + \beta\gamma\cos\beta a}\right). \qquad (3.2)$$

That means, we have to fulfill at the same time the Eqs. (3.1) and (3.2).

© The Author(s) 2015

W. Gräfe, *Quantum Mechanical Models of Metal Surfaces and Nanoparticles*,
SpringerBriefs in Applied Sciences and Technology,
DOI 10.1007/978-3-319-19764-7_3

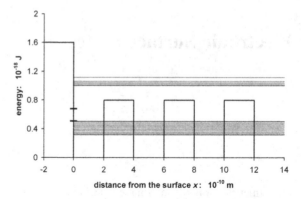

Fig. 3.1 Schematic representation of the run of potential energy in the bulk U and at the surface U^S, of the allowed energy bands (*shaded stripes*), and of the energy levels of Tamm's electronic surface states (*short bars*); Source: J. Mater. Sci. (2013) 48: 2092–2103

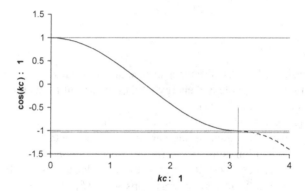

Fig. 3.2 The *thick line* shows the run of cos (kc) and the *dashed line* is the extension cosh (kc) for complex values of k. The *horizontal lines* at the ordinates +1 and −1 are the values of $g(E, U, a, b)$ belonging to the lower and the upper boundaries of the allowed energy band. The *lowest horizontal line* represents a surface state. The *short vertical line* marks the value $kc = \pi$

The numerical calculations have been carried out, as in Chap. 2, for the values $a = b = 2 \times 10^{-10}$ m and for $U = 0.8 \times 10^{-18}$ J (5 eV). And again, the surface potential energy step has the arbitrarily chosen height of $U^S = 1.6 \times 10^{-18}$ J (10 eV).

For these data the fulfillment of Eqs. (2.4), (3.1), and (3.2) is visualized in Figs. 3.2 and 3.3. The condition following from Eq. (3.2) is represented in Fig. 3.3 by the dotted line. Here, the surface state is an *offspring* of the allowed energy band with the energy $E^{So} = 0.506 \times 10^{-18}$ J. A second, not depicted intersection of the solid curve with the dotted line appears at 0.678×10^{-18} J. This is the energy E^{Sa} of an *additional*, upper surface state.

The energies of the two resulting Tamm's electronic surface states are presented in Fig. 3.1 as short horizontal bars.

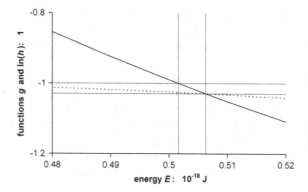

Fig. 3.3 The *thick line* depicts the run of the function $g(E, U, a, b)$ and the *dotted line* illustrates the condition for the existence of a wave function in the surface potential energy step following from Eq. (3.2). The *horizontal line* at -1 is the value of $g(E, U, a, b)$ at the upper boundary of the allowed energy band. The *lower horizontal line* is the value of $g(E, U, a, b)$ for the offspring surface state. The *vertical lines* mark the corresponding energies

Fig. 3.4 Illustration of the ambient influence on the lower surface state; *upper diagram*: The dependence of the surface state energy on the height of the surface potential energy step U^S (The *solid line* represents the top of the allowed energy band.); *lower diagram*: The dependence of the penetration depth of the surface state wave function on the surface potential energy step U^S

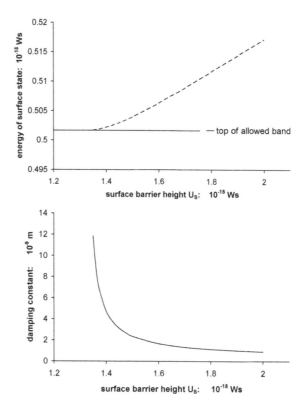

For a height of the potential energy step at the surface $U^S = 1.6 \times 10^{-18}$ J we obtain an attenuation length of the (lower) offspring surface state amounting to $\delta^o = 1.66 \times 10^{-9}$ m. The attenuation length of the wave function for the additional (upper) surface state is $\delta^a = 3.89 \times 10^{-10}$ m.

The diagrams in Fig. 3.4 show the energy of the offspring surface state (the lower surface energy level) versus the height of the surface potential energy step U^S and the corresponding variation of the attenuation length δ^o.

If we reduce the height of the surface potential energy step U^S to nearly 1.3×10^{-18} J the lower surface state disappears.

The parameters of the additional (upper) surface state depend only very weakly on the height of the potential energy barrier at the surface.

Reference

1. Tamm I (1932) Über eine Möglichkeit der Elektronenbindung an Kristalloberflächen. Phys Z Sowj Union 1:733–746

Chapter 4
The Extension of the Kronig–Penney Model by Binding Forces

Abstract For the calculation of attractive forces in the body, a change in the potential energy of the electrons due to a change in the lattice constant of the body is introduced in the model. The force per electron in the periodic potential field of the kind used by Kronig and Penney has been calculated. If the potential energy U in the Schrödinger equation is the sum of three functions each of which depends only on one variable, the Schrödinger equation of the three-dimensional system is separable. If in addition the energy bands are completely filled or completely empty, the problem of the three-dimensional semi-infinite body with a surface can be split into three independent one-dimensional subsystems. Electrons with energy values placed in the energy gaps concerning the motion into the x_k-direction normal to the surface but, at the same time placed in an allowed energy band for the motion into the x_i- or x_j-directions parallel to the surface are localized in a near-surface layer. The electrons localized in the surface bands cause a near-surface local stress with the dimension of a force per area.

Keyword Surface binding forces

In order to calculate attractive forces in the body, a change in the potential energy of the electrons due to a change in the lattice constant of the body must be introduced in the model. Under an expansion in the direction x_i the width of the potential wells a_i shall remain unchanged. However, due to the enlargement of the width of the potential barriers b_i their height U_i shall increase according to Eq. (4.1)

$$U_i = \frac{\partial U_i}{\partial b_i} b_i. \tag{4.1}$$

Such a change of the potential energy due to the expansion of the body is suggested by the change of the potential energy curve for point charges due to an enlargement of the distances between these charges. Both cases are illustrated in

© The Author(s) 2015
W. Gräfe, *Quantum Mechanical Models of Metal Surfaces and Nanoparticles*,
SpringerBriefs in Applied Sciences and Technology,
DOI 10.1007/978-3-319-19764-7_4

Fig. 4.1 Schematic
representation of the change
of the potential energy in a
solid by expansion (*solid line*
equilibrium, *dashed line*
expanded state)

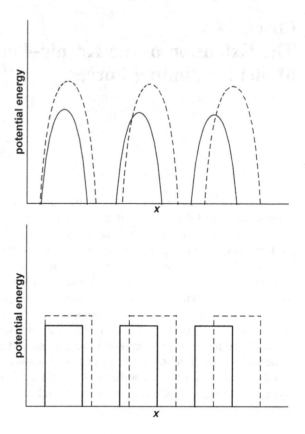

Fig. 4.1. The upper part shows the potential energy caused by point charges and in
the lower part we see the approximation given by Eq. (4.1).

In this way the eigenvalues of the electron energy E on its part adopt the feature
of a potential energy. For the electrons moving in the x_i-direction with a given value
of the wave number k_i, the mechanical force counteracting the expansion of the
body in the x_i-direction is

$$-\frac{\partial E_i}{\partial b_i} = f_i(E_i). \tag{4.2}$$

The differential quotient $\partial E_i/\partial b_i$ follows from Eq. (4.3)

$$-\frac{\partial E_i}{\partial b_i} = \frac{\frac{\partial g_i}{\partial b_i}}{\frac{\partial g_i}{\partial E_i}} = f_i(E_i). \tag{4.3}$$

Fig. 4.2 The energy
dependence of the force per
electron induced by an
expansion of the body
calculated for
$\partial U_i/\partial b_i = 0.4 \times 10^{-8}$ J/m

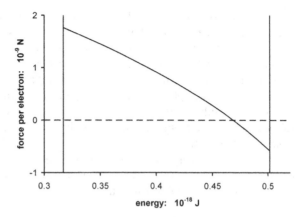

The force per electron in the periodic potential field of the kind used by Kronig
and Penney has been calculated for the values $a_i = b_i = 2 \times 10^{-10}$ m and
$U_i = 0.8 \times 10^{-18}$ J (5 eV). For the rise in the potential energy in the bulk, due to a
stretching of the body, a value of $\partial U_i/\partial b_i = 0.4 \times 10^{-8}$ J/m has been assumed.

Figure 4.2 shows the force per electron $-\partial E_i/\partial b_i = f_i(E_i)$ caused by an
expansion of the body as a function of the energy E_i.

Due to the expansion of a body in the x_i-direction a force in the opposite
direction is generated by all the electrons in the bulk. For this force we make the
ansatz

$$F_i^B\left(E^F\right) = \frac{c_i}{L_i} \int_0^\infty f_i(E)n^B(E)p(E - E^F)\mathrm{d}E. \tag{4.4}$$

The expression $n^B(E)\,\mathrm{d}E$ is the number of states per energy interval from E to
$E + \mathrm{d}E$ for electrons moving in an arbitrary direction in the bulk. The term $p(E-E^F)$
means the Fermi distribution and E^F is the Fermi level in the three-dimensional
body.

If we consider a one-dimensional body with a completely filled energy band, we
obtain the force between two neighboring atoms which is caused by the electrons
moving in the x_i-direction

$$\frac{c_i}{L_i} \int_{E_i^B}^{E_i^T} f_i(E_i)n(E_i)\mathrm{d}E_i. \tag{4.5}$$

The limits of integration E_i^B and E_i^T are the energies at the bottom and the top of
the allowed energy band in the ith subsystem.

In this case the force between two neighboring atoms in the one-dimensional
subsystem caused by one electron in the lowest energy band is 8.2×10^{-10} N.

For a three-dimensional body the contribution of the electrons in the lowest energy band to the cohesive bulk stress is 5.1×10^9 N/m^2.

If the potential energy U in the Schrödinger equation is the sum of three functions each of which depends only on one variable, the Schrödinger equation of the three-dimensional system is separable. If in addition the energy bands are completely filled or completely empty, the problem of the three-dimensional semi-infinite body with a surface can be split into three independent one-dimensional subsystems and the motion of an electron in one direction is independent from its motions in the other directions.

Electrons with energy values placed in the energy gaps concerning the motion into the x_k-direction normal to the surface but, at the same time placed in an allowed energy band for the motion into the x_i- or x_j-directions are localized in a near-surface layer.

The considerations leading to the description of the electron forces in the bulk are also applicable to the electrons, the motion of which is constrained in a layer near the surface. If the body is expanded for instance in the x_i-direction, the electrons localized at the surface give a contribution to a local stress which is concentrated near the surface. For one surface state and the lowest surface energy bands which are completely filled, the corresponding force F_i^S (S-surface) is given by

$$
F_i^S = \frac{c_i}{L_i} \int_0^\infty f_i(E) n^S(E) \mathrm{d}E = \frac{c_i}{L_i} \int_0^\infty \int_{E_i^B}^{E_i^T} \int_{E_j^B}^{E_j^T} f_i(E_i) \{ \delta(E_k - E_k^S) \} n(E_i) n(E_j) \mathrm{d}E_k \mathrm{d}E_i \mathrm{d}E_j
$$

$$
= \frac{c_i L_j}{L_i c_j} \int_{E_i^B}^{E_i^T} f_i(E_i) n(E_i) \mathrm{d}E_i. \tag{4.6}
$$

In Eq. (4.6) the quantity $n^S(E)$ $\mathrm{d}E$ is the number of electrons in the energy range between E and $E + \mathrm{d}E$ moving in a direction parallel to the surface. The quotient F_i^S/L_j has the dimension of a force per length and is the ith component of this quantity in the surface layer parallel to the i- and j-directions. F_i^S/L_j may be regarded as the contribution of the electrons localized in the surface band (ESB) to the tensor component of the surface stress $\sigma_{ii}^{(ESB)}$.

The contribution of the electrons localized in the surface bands to a near-surface local stress with the dimension of a force per area is

$$
s_{ii}^{ESB} = \frac{c_i}{L_i} \frac{1}{c_j} |\psi(x_k)|^2 \int_{E_i^B}^{E_i^T} f_i(E_i) n(E_i) \mathrm{d}E_i. \tag{4.7}
$$

Here, the symbol $\psi(x_k)$ is the wave function of the surface state in the kth subsystem. With a rising distance from the surface into the bulk the linkage forces parallel to the surface decrease. The attenuation length of this near-surface stress is $\delta/2$.

Between $\sigma_{ii}^{(ESB)}$ and $s_{ii}^{(ESB)}$ the relation

$$\sigma_{ii}^{ESB} = \int\limits_{0}^{\infty} s_{ii}^{ESB}(x_k)\,\mathrm{d}x_k \tag{4.8}$$

exists.

Subject to the condition that the equations for the description of the solid and its surface are separable, we obtain for the contribution to the surface stress at $T = 0$ K

$$\frac{F_i^S}{L_j} = -\frac{c_i}{L_i L_j} \int\limits_{0}^{\zeta_i} \frac{\partial E_i}{\partial b_i} n(E_i)\,\mathrm{d}E_i \int\limits_{0}^{\zeta_j} n(E_j)\,\mathrm{d}E_j. \tag{4.9}$$

In Chap. 5 the separation of the three-dimensional body will be considered in detail.

Chapter 5
The Separation of the Semi-infinite Model and the Calculation of the Surface Parameters for the Three-Dimensional body at $T = 0$ K (Regula Falsi of Surface Theory)

Abstract It is demonstrated, how the different ideas "surface energy" and "surface stress" come out from the energy band approximation of a crystalline solid used by Kronig and Penney and from the electronic surface states predicted by Tamm. If the potential energy is separable, we obtain for the chemical potential per particle in the three-dimensional system ζ and the chemical potentials in the subsystems ζ_i the relation $\zeta = \zeta_1 + \zeta_2 + \zeta_3$. By a combination of three one-dimensional models for a solid to a three-dimensional one we obtain with a "regula falsi" (RF) approximate results for the surface charge density $q^{ESB,RF}$, the surface energy $\varphi^{ESB,RF}$, and the surface stress $\sigma_{ii}^{ESB,RF}$. The superfix "ESB" means that we consider here the results for the electrons in the surface bands.

Keyword Separation of three-dimensional model

In the following we will demonstrate, how the different ideas "surface energy" and "surface stress" come out from the energy band approximation of a crystalline solid used by Kronig and Penney [1] and from the electronic surface states predicted by Tamm [2]. (In the text below the "electronic surface states" will be denominated simply "surface states".)

The time-independent Schrödinger equation for the three-dimensional model body is

$$-\frac{\hbar^2}{2m}\Delta\psi + U(x_1, x_2, x_3)\psi = E\psi. \tag{5.1}$$

Kronig and Penney [1] have already mentioned the separability of the time-independent Schrödinger equation in case the potential energy U of the electrons is the sum of three functions each of which depending on one coordinate only.

$$U(x_1, x_2, x_3) = U(x_1) + U(x_2) + U(x_3). \tag{5.2}$$

These three functions shall have the following periodical run

© The Author(s) 2015

W. Gräfe, *Quantum Mechanical Models of Metal Surfaces and Nanoparticles*,
SpringerBriefs in Applied Sciences and Technology,
DOI 10.1007/978-3-319-19764-7_5

$$U(x_i') = 0 \quad \text{for} \quad 0 < x_i' < a_i \tag{5.3a}$$

and

$$U(x_i') = U_i \quad \text{for} \quad a_i \leq x_i' \leq (a_i + b_i) = c_i \tag{5.3b}$$

with $x_i' = x_i - c_i\,(N_i^A - 1)$ for $i = 1, 2, 3$. The quantity N_i^A means the number of an atom arranged in the x_i-direction.

For such a dependence of the potential energy on the space coordinates the Schrödinger Eq. (5.1) may be separated into three wave equations

$$-\frac{\hbar^2}{2m}\Delta\psi_i + U(x_i)\psi_i = E_i\psi_i. \tag{5.4}$$

In this case the wave functions ψ of the Schrödinger Eq. (5.1) are products of the Bloch waves ψ_i for the separate wave Eqs. (5.4)

$$\psi = \psi_1(x_1)\psi_2(x_2)\psi_3(x_3). \tag{5.5}$$

For the energy eigenvalues E of the Schrödinger Eq. (5.1) and the eigenvalues of the separate wave equations E_i it holds the relation

$$E = E_1 + E_2 + E_3. \tag{5.6}$$

That means, we obtain the energy levels in the three-dimensional model simply by adding the contributions of the electron states in the one-dimensional subsystems.

For the electrons moving in the solid in an arbitrary direction the allowed energy levels are again located in separated energy bands. In the three-dimensional model the lower (upper) boundaries of these bands are given by the sum of the lower (upper) boundaries of the allowed energy bands for the motion in one of the directions x_1, x_2, and x_3.

The separability of the Schrödinger equation (5.1) implies the independence of motions of the electrons in three directions x_1, x_2, and x_3. Therefore, the differential of the density of the electrons in the bulk $n^B(E)\mathrm{d}E$ may be substituted by $n(E_1)$ $\mathrm{d}E_1 \cdot n(E_2)\mathrm{d}E_2 \cdot n(E_3)\mathrm{d}E_3$. Here $n(E_i)\mathrm{d}E_i$ is the number of electrons per energy interval from E_i to $E_i + \mathrm{d}E_i$ for a motion into the x_i-direction.

However, these facts do not mean that the three-dimensional model of the solid can be composed by one-dimensional subsystems.

An assembly of the three-dimensional model for the solid surface by three one-dimensional models is only possible if in the formulae of the type

$$\int \{\Phi_1(E_1) + \Phi_2(E_2) + \Phi_3(E_3)\}p(E - E^F)n(E_1)n(E_2)n(E_3)\mathrm{d}E_1\mathrm{d}E_2\mathrm{d}E_3 \tag{5.7}$$

also the Fermi distribution function $p(E-E^F)$ is separable, that is

$$p(E - E^F) = p(E_1 - E_1^F)p(E_2 - E_2^F)p(E_3 - E_3^F). \qquad (5.8)$$

In Eqs. (5.7) and (5.8) the symbols E^F and E_i^F mean the Fermi levels in the three-dimensional system and in the one-dimensional subsystems, respectively.

5.1 The Separability of the Chemical Potential

For a one-dimensional subsystem of the solid, the points of state of the ith one-dimensional subsystem are distributed over the phase space Ω_i spanned by the space coordinates x_i and the impulse coordinates p_i.

For the chemical potential per particle in the ith subsystem ζ_i it holds the relation

$$\zeta_i = -kT\frac{\partial \ln(\Omega_i)}{\partial N_i}. \qquad (5.9)$$

Here ∂N_i means the differential of the particle number in the ith subsystem.

In order to take into consideration that not only the particles arranged along the ith axis are moving into the i-direction, but also the particles in the $N_j \times N_k$ parallel rows, we have to increase the number of points of state by raising Ω_i to the power $N_j N_k$.

We consider each of the enlarged subsystems with the phase space

$$\Omega_i^{N_j N_k} \qquad (5.10)$$

as a separate system in contact with its own thermostat and particle reservoir. In statistics the entity of points of state for such a system is a macro-canonical ensemble.

If we combine the three enlarged subsystems with their separate thermostats and particle reservoirs in one system, we have for the phase space of this combined system

$$\Omega_1^{N_2 N_3}\Omega_2^{N_1 N_3}\Omega_3^{N_1 N_2}. \qquad (5.11)$$

Applying the Taylor series to the exponents in the terms

$$\exp(\ln(\Omega)) \quad \text{and} \quad \exp\left(\ln(\Omega_1^{N_2 N_3}\Omega_2^{N_1 N_3}\Omega_3^{N_1 N_2})\right) \qquad (5.12)$$

we obtain the chemical potentials per particle in the system with particles moving in all three directions ζ and the chemical potentials in the subsystems ζ_i

$$\zeta N = -kT \frac{\partial \ln \Omega}{\partial N} N =$$

$$= -kT \left(\frac{\partial \ln \Omega_1^{N_2 N_3}}{\partial N_1} N_1 + \frac{\partial \ln \Omega_2^{N_1 N_3}}{\partial N_2} N_2 + \frac{\partial \ln \Omega_3^{N_1 N_2}}{\partial N_3} N_3 \right) = (\zeta_1 + \zeta_2 + \zeta_3) N.$$

$$(5.13)$$

As usual in statistics, we have equated here the differential dN of the particle number in the larger part of the whole system, that is the thermostat and the reservoir, with $-N$, the negative number of the particles in the smaller part of the whole system. Correspondingly, we have dealt with the differential of the particle number in the larger part of the subsystems dN_i and the particle number in the smaller part of the subsystems $-N_i$. For the system with the particles moving in all three directions the particle number N is the product of the particle numbers $N_1 N_2 N_3$ in the smaller parts of the subsystems

$$-\partial N \rightarrow N = N_1 N_2 N_3. \qquad (5.14)$$

The symbols N_i used here mean the formulae

$$N_i = \int_0^\infty n(E_i) p(E_i - E_i^F) dE_i = \int_0^\infty n(E_i) p(E_i - \zeta_i) dE_i. \qquad (5.15)$$

According to Eq. (5.13) the chemical potential per particle ζ in the system with particles moving in all three directions is the sum of the chemical potentials per particle in the subsystems $\zeta_1 + \zeta_2 + \zeta_3$.

Now we consider a canonical ensemble from the view point in quantum statistics.

The quantity N_i means the number of the particles in the ith one-dimensional subsystem and $N_j N_k$ is the number of all the subsystems extended in the x_i-direction.

Let Z_i be the partition function of the canonical ensemble for the ith one-dimensional system. If we combine for each index i the $N_j N_k$ ensembles of the one-dimensional systems to an ensemble for the three-dimensional body, we obtain for the partition function Z

$$Z = Z_1^{N_2 N_3} Z_2^{N_1 N_3} Z_3^{N_1 N_2}. \qquad (5.16)$$

In quantum statistics the Helmholtz free energy F is defined as

$$F = -kT \ln(Z) = -kT \ln \left(Z_1^{N_2 N_3} Z_2^{N_1 N_3} Z_3^{N_1 N_2} \right). \qquad (5.17)$$

For a canonical ensemble the chemical potential ζ is

$$\zeta = \frac{\partial F}{\partial N} = -kT\left(\frac{\partial \ln Z}{\partial N}\right) = -kT\left(\frac{\partial \ln Z_1^{N_2 N_3}}{N_2 N_3 \partial N_1} + \frac{\partial \ln Z_2^{N_1 N_3}}{N_1 N_3 \partial N_2} + \frac{\partial \ln Z_3^{N_1 N_2}}{N_1 N_2 \partial N_3}\right)$$
$$= \zeta_1 + \zeta_2 + \zeta_3. \tag{5.18}$$

Again, the terms ∂N and ∂N_i denote the changes of the particle (electron) numbers in the systems under consideration.

For the macro-canonical thermodynamic potential $\Omega'(T,V,\zeta) = -kT\ln(Z)$ the quantity ζ is a natural variable. Because the effects of the three reservoirs connected with the three subsystem are tantamount we have $\zeta = \zeta_1 + \zeta_2 + \zeta_3$.

In a system free of electrical fields the chemical potential ζ and the Fermi level E^F are identical.

5.2 The Separability of the Fermi Distribution Function

According to Eq. (5.6), the Fermi distribution function for the system with particles moving in all three directions takes the form

$$p(E - \zeta) = \frac{1}{1 + e^{\frac{E_1 + E_2 + E_3 - (\zeta_1 + \zeta_2 + \zeta_3)}{kT}}}. \tag{5.19}$$

On the other hand, the Fermi distribution function for the ith partial system is

$$p(E_i - \zeta_i) = \frac{1}{1 + e^{\frac{E_i - \zeta_i}{kT}}}. \tag{5.20}$$

On the conditions

$$e^{\frac{E_i - \zeta_i}{kT}} \gg 1. \tag{5.21}$$

the Fermi distribution function, Eq. (5.20), may be approximated by

$$p(E_i - \zeta_i) = e^{-\frac{E_i - \zeta_i}{kT}} \tag{5.22}$$

and the occupation of the electron levels in the combined system is governed by

$$p(E - \zeta) = e^{-\frac{E_1 + E_2 + E_3 - (\zeta_1 + \zeta_2 + \zeta_3)}{kT}} = p(E_1 - \zeta_1)p(E_2 - \zeta_2)p(E_3 - \zeta_3). \tag{5.23}$$

Correspondingly, for

$$e^{\frac{E_i - \zeta_i}{kT}} \ll 1 \tag{5.24}$$

we obtain

$$1 - p(E - \zeta) = [1 - p(E_1 - \zeta_1)][1 - p(E_2 - \zeta_2)][1 - p(E_3 - \zeta_3)]. \qquad (5.25)$$

5.3 The Calculation of Surface Energy, Surface Stress, and Surface Charge at $T = 0$ K (Regula Falsi of Surface Theory)

Let the body have a plane surface at $x_k = 0$ and be extended in the x_k-direction until $+\infty$. In the x_i- and x_j-directions the body is unlimited.

The electrons localized at the surface with energy values placed in the energy gaps concerning the motion into the x_k-direction, but at the same time placed in the allowed energy bands concerning a motion into the x_i- or x_j-direction, are moving parallel to the surface in a near-surface layer.

In case an additional surface state Sa is generated due to the production of a new surface, we replace the distribution of the energy levels $n(E)dE$ by

$$n(E)dE \rightarrow \left(n(E_k) + \delta(E_k - E^{\mathrm{Sa}})\right)n(E_i)n(E_j)dE_kdE_idE_j. \qquad (5.26)$$

Given that the conditions for a separation of the Fermi distribution is fulfilled we obtain for the total energy of the system

$$
\begin{aligned}
E = \int\limits_0^\infty n(E)Ep(E - \zeta)dE \rightarrow \\
\approx \int\limits_0^\infty \int\limits_0^\infty \int\limits_0^\infty \left(n(E_k) + \delta(E_k - E^{\mathrm{Sa}})\right)n(E_i)n(E_j)(E_k + E_i + E_j) \\
\times p(E_k - \zeta_k)p(E_i - \zeta_i)p(E_j - \zeta_j)dE_kdE_idE_j
\end{aligned}
\qquad (5.27)
$$

and have

$$
\begin{aligned}
E^{RF} = \int\limits_0^\infty \int\limits_0^\infty \int\limits_0^\infty n(E_k)n(E_i)n(E_j)(E_k + E_i + E_j)p(E_k - \zeta_k)p(E_i - \zeta_i)p(E_j - \zeta_j)dE_kdE_idE_j \\
+ \int\limits_0^\infty \int\limits_0^\infty \int\limits_0^\infty \delta(E_k - E^{\mathrm{Sa}})n(E_i)n(E_j)(E_k + E_i + E_j)p(E_k - \zeta_k)p(E_i - \zeta_i)p(E_j - \zeta_j)dE_kdE_idE_j.
\end{aligned}
$$

$$(5.28)$$

The superscript ^{RF} means regula falsi. The first term in Eq. (5.28) is the energy of the electrons moving in the bulk whereas the second term is the energy of the electrons captured at the surface. E^{Sa} denotes the energy of the additional surface state. The division of the second term in Eq. (5.28) by the surface area results in the surface energy φ^{ESB} (^{ESB} means the electrons in the surface band).

$$
\varphi^{ESB,RF} = \frac{1}{L_i L_j} \left\{ E^{Sa} p(E^{Sa} - \zeta_k) \int_0^\infty n(E_i) p(E_i - \zeta_i) dE_i \int_0^\infty n(E_j) p(E_j - \zeta_j) dE_j \right.
$$

$$
+ p(E^{Sa} - \zeta_k) \int_0^\infty n(E_i) E_i p(E_i - \zeta_i) dE_i \int_0^\infty n(E_j) p(E_j - \zeta_j) dE_j
$$

$$
\left. + p(E^{Sa} - \zeta_k) \int_0^\infty n(E_i) p(E_i - \zeta_i) dE_i \int_0^\infty n(E_j) E_j p(E_j - \zeta_j) dE_j \right\}.
$$

$$(5.29)$$

For temperatures near the absolute zero the Fermi distributions degenerate into step functions with the step at $E_i - \zeta_i = 0$. That means, with the exclusion of the step either the condition $\exp((E_i - \zeta_i)/kT) \ll 1$ or $\exp((E_i - \zeta_i)/kT) \gg 1$ holds.

From Eq. (5.29) we obtain for the temperature $T = 0$ K in the case $\zeta_k \geq E^{Sa}$ the contribution of the surface state Sa to the surface energy of the electrons in the surface band

$$
\varphi^{ESB,RF} = \frac{1}{L_i L_j} \left\{ E^{Sa} \int_0^{\zeta_i} n(E_i) dE_i \int_0^{\zeta_j} n(E_j) dE_j + \int_0^{\zeta_i} n(E_i) E_i dE_i \int_0^{\zeta_j} n(E_j) dE_j \right.
$$

$$
\left. + \int_0^{\zeta_i} n(E_i) dE_i \int_0^{\zeta_j} n(E_j) E_j dE_j \right\}.
$$

$$(5.30)$$

For $\zeta_k < E^{Sa}$ the contribution of the additional surface state to the surface energy at the temperature $T = 0$ K is zero.

In the following we take into consideration

$$
\partial \varepsilon_{ii} = \frac{\partial L_i}{L_i} = \frac{\partial b_i}{c_i}. \tag{5.31}
$$

The energy of the surface state E^{Sa} is determined by parameters belonging to the x_k-system only. Therefore, we have $\partial E^{Sa}/\partial \varepsilon_{ii} = \partial E^{Sa} \partial \varepsilon_{jj} = 0$. Furthermore, we suppose that during the stretching of the specimen the Fermi level remains constant. Then, for $\zeta_k \geq E^{Sa}$, we obtain for the derivative of the surface energy with respect to ε_{ii} at the temperature $T = 0$ K

$$\frac{\partial \varphi^{ESB,RF}}{\partial \varepsilon_{ii}} = -\varphi^{ESB,RF}\delta_{ii} + \frac{1}{L_iL_j}\left\{ E^{Sa} \int\limits_0^{\zeta_i} \frac{\partial n(E_i)}{\partial \varepsilon_{ii}}dE_i \int\limits_0^{\zeta_j} n(E_j)dE_j \right.$$

$$+ \left(\int\limits_0^{\zeta_i} \frac{\partial n(E_i)}{\partial \varepsilon_{ii}}E_idE_i + \int\limits_0^{\zeta_i} n(E_i)\frac{\partial E_i}{\partial \varepsilon_{ii}}dE_i \right) \int\limits_0^{\zeta_j} n(E_j)dE_j \qquad (5.32)$$

$$\left. + \int\limits_0^{\zeta_i} \frac{\partial n(E_i)}{\partial \varepsilon_{ii}}dE_i \int\limits_0^{\zeta_j} n(E_j)E_jdE_j \right\}.$$

On the same conditions the surface stress component σ_{ii} for the model used here amounts to

$$\sigma_{ii}^{ESB,RF} = \varphi^{ESB,RF}\delta_{ii} + \frac{\partial \varphi^{ESB,RF}}{\partial \varepsilon_{ii}} = \frac{1}{L_iL_j}\left\{ E^{Sa} \int\limits_0^{\zeta_i} \frac{\partial n(E_i)}{\partial \varepsilon_{ii}}dE_i \int\limits_0^{\zeta_j} n(E_j)dE_j \right.$$

$$+ \left(\int\limits_0^{\zeta_i} \frac{\partial n(E_i)}{\partial \varepsilon_{ii}}E_idE_i + \int\limits_0^{\zeta_i} n(E_i)\frac{\partial E_i}{\partial \varepsilon_{ii}}dE_i \right) \int\limits_0^{\zeta_j} n(E_j)dE_j \qquad (5.33)$$

$$\left. + \int\limits_0^{\zeta_i} \frac{\partial n(E_i)}{\partial \varepsilon_{ii}}dE_i \int\limits_0^{\zeta_j} n(E_j)E_jdE_j \right\}.$$

In contrast to Eq. (4.9) for the surface stress

$$\frac{F_i^S}{L_j} = -\frac{c_i}{L_iL_j} \int\limits_0^{\zeta_i} \frac{\partial E_i}{\partial b_i}n(E_i)dE_i \int\limits_0^{\zeta_j} n(E_j)dE_j \qquad (4.9)$$

in Eq. (5.33) additional terms appear which contain the derivation $\partial n/\partial \varepsilon = c \cdot \partial n/\partial b$.

In Fig. 5.1 we compare the values for the surface stress at $T = 0$ K calculated with the Eqs. (5.33) and (4.9).

We see that in the scale of Fig. 5.1 both curves coincide. The difference between σ_{ii} and F_i^S/L_j in relation to the maxima of the particular values is smaller than 10^{-34}!

The Eq. (4.9) is the result of a simple and self-evident deduction. It is important that this derivation has been accomplished without any recourse to the surface energy and the formula of Herring [3] in Eq. (1.3).

Fig. 5.1 Comparison of the results for $\sigma_{ii}^{\text{ESB,RF}}(\zeta_i)$ (*solid line*) with the data of $F_i^S/L_j(\zeta_i)$ (*squares*) at $T = 0$ K according to the Eqs. (5.33) and (4.9)

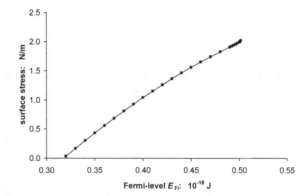

The surface charge density of the electrons in the surface band q^{ESB} is given by

$$
q^{\text{ESB}} = -\frac{e}{L_i L_j} \int_0^\infty n^S(E) p(E - \zeta) dE \approx q^{\text{ESB,RF}}
$$

$$
= -\frac{e}{L_i L_j} p(E^{\text{Sa}} - \zeta_k) \int_0^\infty n(E_i) p(E_i - \zeta_i) dE_i \int_0^\infty n(E_j) p(E_j - \zeta_j) dE_j
$$

(5.34)

The symbol e means the absolute value of the electron charge. If the condition $\zeta_k \geq E^{\text{Sa}}$ holds, it follows from Eq. (5.34) for the surface charge density at the temperature $T = 0$ K

$$
q^{\text{ESB,RF}} = -\frac{e}{L_i L_j} \int_0^{\zeta_i} n(E_i) dE_i \int_0^{\zeta_j} n(E_j) dE_j.
$$

(5.35)

For $\zeta_k < E^{\text{Sa}}$ the contribution of the surface state Sa at $T = 0$ K to the surface charge density q is zero.

As discussed in Chap. 3, a surface state may be the offspring from an energy band. In this case, the number of the energy levels in the energy band the offspring stems from is reduced by 1.

If $n'(E_k) dE_k$ is the reduced number of energy levels in the energy range from E_k to $E_k + dE_k$, we have

$$
\int_{E_k^B}^{E_k^T} n'(E_k) dE_k = \int_{E_k^B}^{E_k^T} n(E_k) dE_k - 1.
$$

(5.36)

Therefore, for an offspring surface state So we have to use instead of Eq. (5.26)

$$n(E)dE \rightarrow \left(n'(E_k) + \delta\left(E_k - E^{So}\right)\right)n(E_i)n(E_j)dE_k dE_i dE_j. \tag{5.26a}$$

Nevertheless, for the surface energy caused by an offspring surface state So we obtain in this case the similar formula

$$
\varphi^{\text{ESB,RF}} = \frac{1}{L_i L_j} \left\{ E^{So} p(E^{So} - \zeta_k) \int_0^\infty n(E_i)p(E_i - \zeta_i)dE_i \int_0^\infty n(E_j)p(E_j - \zeta_j)dE_j \right.
$$

$$
+ p(E^{So} - \zeta_k) \int_0^\infty n(E_i)E_i p(E_i - \zeta_i)dE_i \int_0^\infty n(E_j)p(E_j - \zeta_j)dE_j
$$

$$
\left. + p(E^{So} - \zeta_k) \int_0^\infty n(E_i)p(E_i - \zeta_i)dE_i \int_0^\infty n(E_j)E_j p(E_j - \zeta_j)dE_j \right\}.
$$

$$\tag{5.29a}$$

The appearance of the offspring surface state is accompanied with a reduction of the binding forces in the bulk.

The term $n'(E_k)$ is considered in more detail in the discussion of the surface free energy in Chap. 7.

With the combination of three one-dimensional models for a solid to a three-dimensional one the surface charge density $q^{\text{ESB,RF}}$, the surface energy $\varphi^{\text{ESB,RF}}$, and the surface stress $\sigma_{ii}^{\text{ESB,RF}}$ have been calculated for $T = 0$ K.

The graphs for $\varphi^{\text{ESB,RF}}(\zeta)$ and $q^{\text{ESB,RF}}(\zeta)$ are presented in Figs. 6.1 and 6.3.

References

1. de Kronig RL, Penney WG (1931) Quantum mechanics of electrons in crystal lattices. Proc Roy Soc 130(Ser A):499–513
2. Tamm I (1932) Über eine Möglichkeit der Elektronenbindung an Kristalloberflächen. Phys Z Sowj Union 1:733–746
3. Herring C (1951) Surface tension as a motivation for sintering. In: Kingston WE (ed) The physics of powder metallurgy. McGraw-Hill, New York, pp 165 and 143–180

Chapter 6
The Surface Parameters for the Semi-infinite Three-Dimensional Body at Arbitrary Temperature

Abstract The three-dimensional body shall be semi-infinite and terminated by a surface in the plane $x_k = 0$. For the correct calculation of the surface energy of the electrons in the surface band φ^{ESB} we find a convolution integral. Applying the Herring equation to φ^{ESB} we obtain the surface stress of the electrons in the surface band σ^{ESB}. The surface charge density of the electrons in the surface band q^{ESB} has also been calculated. The comparison of the correct results for the surface parameters with the approximate outcomes obtained with the "regula falsi" in Chap. 5 as functions of chemical potential shows that the data are identical for the completely empty and completely occupied energy bands. The application of the convolution to the model for the semi-infinitely extended body allows the use of the Fermi distribution function for the three-dimensional model without any recourse to the Fermi distributions for the one-dimensional systems as it was needed in Chap. 5. If we depict the surface stress σ^{ESB} as a function of the surface charge density q^{ESB}, there is only a little deviation between the approximate and correct results.

Keyword Convolution in surface theory

Again, the body shall be semi-infinite and terminated by a surface in the plane $x_k = 0$. This surface is physically described by a step of the potential energy $U(x_k)$ of the height U^S with $U^S > U_k$ as schematically represented in Fig. 3.1.

If the conditions, Eqs. (5.18) and (5.21) are fulfilled, we obtain for the total energy of the system

$$E \approx \int_0^\infty \int_0^\infty \int_0^\infty n(E_k)n(E_i)n(E_j)(E_k + E_i + E_j)p(E_k - \zeta_k)p(E_i - \zeta_i)p(E_j - \zeta_j)dE_k dE_i dE_j$$

$$+ \int_0^\infty \int_0^\infty \int_0^\infty \delta(E_k - E^{Sa})n(E_i)n(E_j)(E_k + E_i + E_j)p(E_k - \zeta_k)p(E_i - \zeta_i)p(E_j - \zeta_j)dE_k dE_i dE_j.$$

$$(5.28)$$

© The Author(s) 2015
W. Gräfe, *Quantum Mechanical Models of Metal Surfaces and Nanoparticles*,
SpringerBriefs in Applied Sciences and Technology,
DOI 10.1007/978-3-319-19764-7_6

The first term in Eq. (5.28) is the energy of the electrons moving in the bulk whereas the second term is the energy of the electrons captured at the surface. E^{Sa} denotes the energy of the additional surface state.

Correspondingly, for an offspring surface state (So) we have

$$
E \approx \int_0^\infty \int_0^\infty \int_0^\infty n'(E_k)n(E_i)n(E_j)(E_k + E_i + E_j)p(E_k - \zeta_k)p(E_i - \zeta_i)p(E_j - \zeta_j)dE_k dE_i dE_j
$$

$$
+ \int_0^\infty \int_0^\infty \int_0^\infty \delta(E_k - E^{So})n(E_i)n(E_j)(E_k + E_i + E_j)p(E_k - \zeta_k)p(E_i - \zeta_i)p(E_j - \zeta_j)dE_k dE_i dE_j.
$$

$$
(5.28a)
$$

If the conditions, Eqs. (5.18) and (5.21) are fulfilled, the second term in Eq. (5.28)

$$
\int_0^\infty \int_0^\infty (E^{Sa} + E_i + E_j)n(E_i)n(E_j)p(E^{Sa} - \zeta_k)p(E_i - \zeta_i)p(E_j - \zeta_j)dE_i dE_j \quad (6.1)
$$

is correct but, its range of validity is very restricted. However, the term is qualified to give an allusion to the correct formulae for the calculation of the surface parameters.

That is to say, if we invert the chain of conclusions, we obtain

$$
\int_0^\infty \int_0^\infty (E^{Sa} + E_i + E_j)n(E_i)n(E_j)p(E^{Sa} + E_i + E_j - \zeta)dE_i dE_j. \quad (6.2)
$$

With $E_{i+j} = E_i + E_j$ the term in Eq. (6.2) can be transformed into a convolution integral.

$$
\int_0^\infty p(E^{Sa} + E_{i+j} - \zeta) \int_0^\infty (E^{Sa} + E_{i+j})n(E_i)n(E_{i+j} - E_i)dE_i dE_{i+j}. \quad (6.3)
$$

The energy of the electrons in the surface band $(^{ESB})$ belonging to E_k^S, that means, e.g., the integrals in Eqs. (6.1)–(6.3), dealt by the area of the surface under consideration $L_i L_j$, is the surface energy φ^{ESB} in Eq. (6.4).

$$\varphi^{ESB,Sa} = \frac{1}{L_i L_j} \int_0^\infty p(E^{Sa} + E_{i+j} - \zeta) \int_0^\infty (E^{Sa} + E_{i+j})n(E_i)n(E_{i+j} - E_i)dE_i dE_{i+j}.$$

(6.4)

Applying the Herring equation, Eq. (1.3), to Eq. (6.4) we obtain for the surface stress,

$$\sigma_{ii}^{ESB,Sa} = \varphi^{ESB,Sa}\delta_{ii} + \frac{\partial \varphi^{ESB,Sa}}{\partial \varepsilon_{ii}}$$

$$= \frac{1}{L_i L_j} \frac{\partial}{\partial \varepsilon_{ii}} \int_0^\infty p(E^{Sa} + E_{i+j} - \zeta) \int_0^\infty (E^{Sa} + E_{i+j})n(E_i)n(E_{i+j} - E_i)dE_i dE_{i+j}$$

$$= \frac{1}{L_i L_j} \left\{ \int_0^\infty p(E^{Sa} + E_{i+j} - \zeta) \left(\int_0^\infty \frac{\partial E_i}{\partial \varepsilon_{ii}} n(E_i)n(E_{i+j} - E_i)dE_i \right. \right.$$

$$\left. \left. + \int_0^\infty (E^{Sa} + E_{i+j})\frac{\partial n(E_i)}{\partial \varepsilon_{ii}} n(E_{i+j} - E_i)dE_i \right) dE_{i+j} \right\}.$$

(6.5)

It remains in the question, why no derivations of the Fermi functions appear in the Eq. (6.5) for the calculations of the surface stress. The reason is as follows:

The meaning of the symbol E is ambiguous. On the one hand, the letter E stands for the eigenvalues of the Schrödinger equation and on the other hand it is an integration variable. Therefore, we could label the different notions of the energy E with E^e for the eigenvalue and by η for the integration variable. The variable η is defined in the range from $-\infty$ to $+\infty$ but in contrast, the eigenvalues E^e exist only in the allowed energy bands and the surface states.

For the derivation of the integral with respect to the parameter ε in Eq. (6.5) we have to apply the Leibniz rule

$$\frac{d}{d\varepsilon} \int_a^b f(x,\varepsilon)dx = \int_a^b \frac{\partial f(x,\varepsilon)}{\partial \varepsilon}dx = \int_{E^B}^{E^T} \frac{\partial(E^e(\eta,\varepsilon)n(\eta,\varepsilon)p(\eta,\zeta))}{\partial \varepsilon}d\eta, \qquad (6.6)$$

see Rothe [3] and Smirnow [4]. That means, we have to carry out the partial derivation of the integrand on the right-hand side of Eq. (6.6). The eigenvalue E^e depends on ε whereas the (independent) integration variable η does not. That means, we have to observe in the derivation of φ with respect to ε_{ii} the dependencies $E_i^e = E_i^e(\eta_i, \varepsilon_{ii})$, $L_i = L_i(\varepsilon_{ii})$, and $n_i = n_i(\eta_i, \varepsilon_{ii})$. However, the Fermi distribution function $p(\eta, \zeta)$ does not explicitly depend on ε_{ii}.

The surface charge density of the electrons in the surface band $q^{ESB,Sa}$ is given by

$$q^{ESB,Sa} = -\frac{e}{L_i L_j} \int_0^\infty p(E^{Sa} + E_{i+j} - \zeta) \int_0^\infty n(E_i)n(E_{i+j} - E_i)dE_i dE_{i+j}. \qquad (6.7)$$

For the results presented in the Figs. 6.1–6.4 the computations have been restricted to the lowest energy band and the lowest surface state.

The Fig. 6.1 shows the surface energy $\varphi^{ESB,Sa}$ at $T = 300$ K as a function of the chemical potential ζ for the three-dimensional model according to Eq. (6.4) and the approximate values as they follow from the corresponding formula (5.30) for the conditions in Eqs. (5.18) and (5.21).

In Fig. 6.2 we compare the surface stress $\sigma^{ESB,Sa}$ at $T = 300$ K depicted versus the chemical potential ζ for the three-dimensional model according to Eq. (6.5) with the approximate quantities for $T = 0$ K calculated with the corresponding Eq. (5.33).

The Fig. 6.3 illustrates the dependence of the surface charge density $q^{ESB,Sa}$ at $T = 300$ K on the chemical potential ζ for the three-dimensional model according to Eq. (6.7) and the quantities for $T = 0$ K following from the corresponding Eq. (5.34).

The principal existence of a cohesive stress with the dimension of a force per area which is concentrated in a near-surface layer has been demonstrated in Eq. (4.7). For the surface stress component considered in Eq. (6.5), the near-surface local stress is

$$s_{ii}^{ESB,Sa} = |\psi(x_k)|^2 \sigma_{ii}^{ESB,Sa} \qquad (6.8)$$

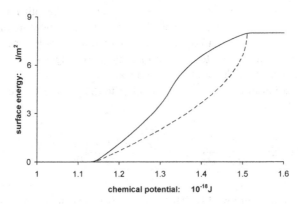

Fig. 6.1 Surface energy $\varphi^{ESB,Sa}$ for $T = 300$ K as a function of the chemical potential ζ calculated with the convolution integral according to Eq. (6.4) (*solid line*) and the approximate values of the surface energy $\varphi^{ESB,RF}$ resulting from the corresponding Eq. (5.30) for $T = 0$ K (*dashed line*); *Source* J. Mater. Sci. (2014) 49: 558–561

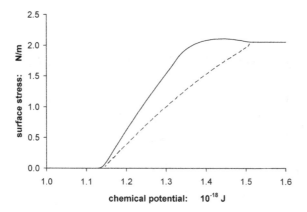

Fig. 6.2 Surface stress $\sigma^{ESB,Sa}$ at $T = 300$ K as a function of the chemical potential ζ calculated with the convolution integral in Eq. (6.5) (*solid line*) and the surface stress $\sigma^{ESB,RF}$ calculated with the approximating Eq. (5.33) for $T = 0$ K (*dashed line*); *Source* J. Mater. Sci. (2014) 49: 558–561

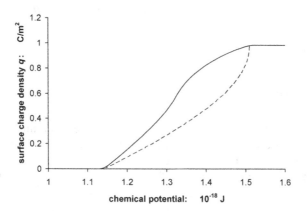

Fig. 6.3 Surface charge density $q^{ESB,Sa}$ at $T = 300$ K as a function of the chemical potential ζ calculated with the convolution integral in Eq. (6.7) (*solid line*) and the corresponding values $q^{ESB,RF}$ for $T = 0$ K resulting from the accordant Eq. (5.34) (*dashed line*); *Source* J. Mater. Sci. (2014) 49: 558–561

with the wave function of the surface state under consideration $\psi(x_k)$.

For the surface state So the corresponding equations can be deduced.

$$\varphi^{ESB,So} = \frac{1}{L_i L_j} \int_0^\infty p(E^{So} + E_{i+j} - \zeta) \int_0^\infty (E^{So} + E_{i+j}) n(E_i) n(E_{i+j} - E_i) dE_i dE_{i+j}$$

$$(6.4a)$$

and

$$
\begin{aligned}
\sigma_{ii}^{ESB,So} &= \frac{1}{L_i L_j} \frac{\partial}{\partial \varepsilon_{ii}} \int_0^\infty p(E^{So} + E_{i+j} - \zeta) \int_0^\infty (E^{So} + E_{i+j}) n(E_i) n(E_{i+j} - E_i) dE_i dE_{i+j} \\
&= \frac{1}{L_i L_j} \left\{ \int_0^\infty p(E^{So} + E_{i+j} - \zeta) \left(\int_0^\infty \frac{\partial E_i}{\partial \varepsilon_{ii}} n(E_i) n(E_{i+j} - E_i) dE_i \right. \right. \\
&\quad + \left. \left. \int_0^\infty (E^{So} + E_{i+j}) \frac{\partial n(E_i)}{\partial \varepsilon_{ii}} n(E_{i+j} - E_i) dE_i \right) dE_{i+j} \right\}.
\end{aligned}
$$

$$(6.5a)$$

Correspondingly, the surface charge density of the electrons in the surface band $q^{ESB,So}$ is given by

$$
q^{ESB,So} = -\frac{e}{L_i L_j} \int_0^\infty p(E^{So} + E_{i+j} - \zeta) \int_0^\infty n(E_i) n(E_{i+j} - E_i) dE_i dE_{i+j}. \qquad (6.7a)
$$

6.1 Discussion

The application of the convolution to the model of the semi-infinitely extended body allows the use of the Fermi distribution function for the three-dimensional model without any recourse to the Fermi distributions for the one-dimensional systems as it was needed in Chap. 5. Therefore, it renders possible to compute correctly the surface charge density q^{ESB}, the surface energy φ^{ESB}, and the surface stress σ^{ESB}.

The run of the surface stress in Fig. 6.2 shows a maximum. This feature is well known from electrocapillarity as it has been discussed in relation with the Eq. (1.1). For the sake of illustration, the Fig. 6.4 shows the surface stress depicted as a function of the surface charge density. For comparison, the corresponding functional dependence is presented for the approximate data obtained with the "regula falsi of surface theory" in Chap. 5. (In an equivalent diagram for the surface energy versus the surface charge density the graph resulting from the convolution and the graph for the approximate data coincide almost completely.)

The surplus near-surface local stress calculated in this chapter causes a compression in all the bulk.

In Chap. 3 the attenuation length of the surface state wave function has been correctly determined. Knowing this quantity one can calculate the near-surface stress as it has also been discussed in Chap. 3.

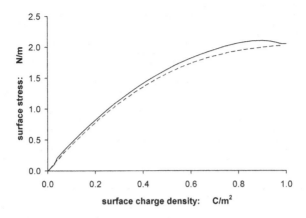

Fig. 6.4 Surface stress versus the surface charge density calculated for $T = 300$ K with convolution integrals according to Eqs. (6.5) and (6.7) (*solid line*) and the same function for the approximate data $\sigma^{ESB,RF}$ and $q^{ESB,RF}$ reckoned for $T = 0$ K in Chap. 5 with the "regula falsi of surface theory" (*dashed line*); *Source* J. Mater. Sci. (2014) 49: 558–561

The infinitely extended model of Kronig and Penney [1] and the semi-infinite model of Tamm [2] are appropriate for a discussion of the effects caused by the electrons in the surface states.

In the foregoing text, we have considered the theoretical relations between the near-surface stress and the common surface parameters, that is to say the surface charge density q^{ESB}, the surface energy φ^{ESB}, and the surface stress σ^{ESB}.

I assume that the near-surface stress is the cause of the fatigue limit playing an important role in materials testing. This is explained in more detail in Chap. 14.

It is the aim of the present explanations to demonstrate the way for a principally correct numerical calculation of the surface parameters at an arbitrary temperature and for an arbitrary position of the Fermi level in the three-dimensional model.

In contrast to the numerical results of the first principles calculations achieved in the literature which are mentioned in Chap. 1, in the present text only an assessment is aspired for the quantitative contributions of the electrons in the Tamm's surface states to the surface parameters under consideration.

References

1. de Kronig RL, Penney WG (1931) Quantum mechanics of electrons in crystal lattices. Proc. Roy. Soc. Ser. A 130:499–513
2. Tamm I (1932) Über eine Möglichkeit der Elektronenbindung an Kristalloberflächen. Phys. Z. Sowj. Union 1:733–746
3. Rothe R (1954) Höhere Mathematik für Mathematiker, Physiker, Ingenieure, Teil II, B. Teubner Verlagsgesellschaft Leipzig, S. 143
4. Smirnow WI (1955) Lehrgang der höheren Mathematik, Teil III,2, VEB Deutscher Verlag der Wissenschaften Berlin, 271

Chapter 7
The Surface Free Energy φ and the Point of Zero Charge Determined for the Semi-infinite Model

Abstract The surface free energy φ of a semi-infinite three-dimensional body has been calculated with the convolution integrals. Applying the Herring's equation to the surface free energy, the surface stress σ has been determined. The differences between the surface energy of the electrons in the surface band φ^{ESB} and the surface free energy φ are discussed as well as the different meanings of the surface stress caused by the electrons in the surface band σ^{ESB} and the surface stress σ resulting from the application of Herring's equation to the surface free energy φ. Also with the convolution integrals, the total surface charge density has been determined. The point of the vanishing total charge is assigned to the point of zero charge (PZC). The contribution of the electrostatic repulsion between the electrons in the surface bands to the surface energy has been discussed sketchily.

Keywords Surface free energy · Point of zero charge

7.1 Electron Transitions from the Bulk into the Surface and the Contribution to the Surface Free Energy φ^{Tr}

At first, we consider a rough estimation of the surface free energy for completely occupied and completely empty energy bands only. In the subsystem for the motion of electrons in the direction normal to the surface, we select as the starting level for an electron transition from the bulk into the surface the mean energy $\bar{E}_k$. The target level is the surface state E_k^S.

For the values of the lattice parameters a and b, we use as before $a = b = 2 \times 10^{-10}$ m. Let in following considerations, the potential barriers in the bulk and at the surface also amount to $U_i = 0.8 \times 10^{-18}$ J and $U^S = 1.6 \times 10^{-18}$ J, respectively, we obtain

© The Author(s) 2015
W. Gräfe, *Quantum Mechanical Models of Metal Surfaces and Nanoparticles*,
SpringerBriefs in Applied Sciences and Technology,
DOI 10.1007/978-3-319-19764-7_7

the number of atoms per length 2.5×10^9/m,
the mean energy of the electrons in the lower energy band $\bar{E}_k = 0.395 \times 10^{-18}$ J,
and
the energy of the lower surface state $E_k^{So} = 0.506 \times 10^{-18}$ J.

With these values, we find the contributions of the electron transitions to the surface free energy

$$\varphi^{TR} = (E_k^S - \bar{E}_k)N^2 \approx 0.7 \text{ J/m}^2. \tag{7.1}$$

For a more detailed consideration, we split the solid into a series of lattice planes parallel to the surface and assign to each atom in the layer the energy $\bar{E}_k(\zeta)$ with

$$\bar{E}_k(\zeta) = \frac{\int\limits_0^\infty E_k n(E_k) \dfrac{1}{1+e^{\frac{E_k-\zeta}{kT}}} dE_k}{\int\limits_0^\infty n(E_k) \dfrac{1}{1+e^{\frac{E_k-\zeta}{kT}}} dE_k}. \tag{7.2}$$

Now we consider in Eq. (7.3) the difference between the energy of the electrons in the surface band φ^{ESB} and the energy of the electrons in one of the mentioned layers parallel to the surface. The occupation of the surface band and of the energy band in the lattice planes is controlled by the same position of the chemical potential ζ.

$$\varphi^{TR} = \frac{1}{L_i L_j} \left\{ \int\limits_0^\infty (E_k^S + E_{i+j})n(E_{i+j}) \frac{1}{1+e^{\frac{E_k^S + E_{i+j} - \zeta}{kT}}} dE_{i+j} \right\}$$
$$- \frac{1}{L_i L_j} \left\{ \int\limits_0^\infty (\bar{E}_k(\zeta) + E_{i+j})n(E_{i+j}) \frac{1}{1+e^{\frac{\bar{E}_k(\zeta) + E_{i+j} - \zeta}{kT}}} dE_{i+j} \right\}. \tag{7.3}$$

Here, the choice of $\bar{E}_k(\zeta)$ as the starting level is an arbitrary act.

In the result of a reversible process, a free energy reaches its minimum value. For the electron transitions from the bulk into the surface, the contribution to the surface free energy φ^{TR} equals the lowest values by electron transitions from the top level of the energy band E_k^T to the surface state E_k^S. With a similar argumentation as above for the starting level $\bar{E}_k(\zeta)$, we obtain now for the surface free energy.

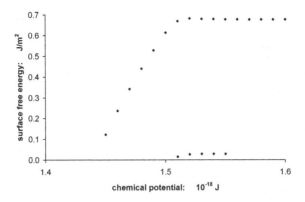

Fig. 7.1 The contributions to the surface free energy φ^{TR} resulting from the electron transitions from the bulk into the lower surface band of the lower surface state at $T = 300$ K were calculated with the convolution integrals in Eqs. (7.3) and (7.4) and are depicted as functions of the chemical potential ζ. The starting levels for the electron transitions into the lower surface state are E_k^T (*lower curve*) and $\bar{E}_k(\zeta)$ (*higher curve*)

$$
\varphi^{TR} = \varphi^{ESB} - \frac{1}{L_i L_j} \left\{ \int_0^\infty (E_k^T + E_{i+j}) n(E_{i+j}) \frac{1}{1 + e^{\frac{E_k^T + E_{i+j} - \zeta}{kT}}} dE_{i+j} \right\}
$$

$$
= \frac{1}{L_i L_j} \left\{ \int_0^\infty (E_k^S + E_{i+j}) n(E_{i+j}) \frac{1}{1 + e^{\frac{E_k^S + E_{i+j} - \zeta}{kT}}} dE_{i+j} \right\} \tag{7.4}
$$

$$
- \frac{1}{L_i L_j} \left\{ \int_0^\infty (E_k^T + E_{i+j}) n(E_{i+j}) \frac{1}{1 + e^{\frac{E_k^T + E_{i+j} - \zeta}{kT}}} dE_{i+j} \right\}.
$$

The contributions of the electron transitions to the surface free energy φ^{TR} at $T = 300$ K for the electrons in the lower surface band of the lower surface state have been calculated with the convolution integrals in Eqs. (7.3) and (7.4) for both the starting levels of the electron transitions, $\bar{E}_k(\zeta)$ and E_k^T. The results are depicted in Fig. 7.1 versus the chemical potential ζ.

For the starting level $\bar{E}_k(\zeta)$, the saturation value of the contribution to the surface free energy φ^{TR} amounts here, as in Eq. (7.1), also to nearly 0.7 J/m^2.

For the transition from starting level E_k^T belonging to the lower energy band in the bulk to the additional upper surface state as the target level, the contribution to the surface free energy φ^{TR} is presented as a function of the chemical potential ζ in Fig. 7.2.

The contribution to the surface free energy φ^{TR} of additional electrons transferred from the outside into the body by conduction is equal to φ^{ESB}!

Now we apply the Herring's equation (1.3) to φ^{TR} and obtain

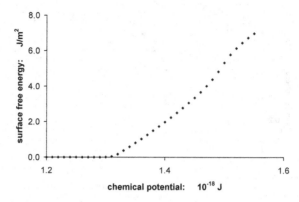

Fig. 7.2 Contribution to the surface free energy φ^{TR} of the electron transitions from E_k^T into the lower energy band of the upper surface state at $T = 300$ K calculated with the convolution integral in Eq. (7.4) as a function of the chemical potential ζ

$$
\sigma_{ii}^{TR,Sa} = \varphi^{TR,Sa} + \frac{\partial}{\partial \varepsilon_{ii}} \varphi^{TR,Sa}
$$

$$
= \sigma^{ESB} - \frac{1}{L_i L_j} \frac{\partial}{\partial \varepsilon_{ii}} \left\{ \int_0^\infty (E_k^T + E_{i+j}) n(E_{i+j}) \frac{1}{1 + e^{\frac{E_k^T + E_{i+j} - \zeta}{kT}}} dE_{i+j} \right\}. \tag{7.5}
$$

The term following the quantity σ^{ESB} in Eq. (7.5) is related only to the energy levels in the bulk.

Equation (7.5) describes the enhancement of the near-surface cohesive forces due to the generation of the surface. In contrast to $\sigma^{TR,Sa}$ in the surface stress σ^{ESB} are included also the forces which have existed between the atoms in the bulk before the generation of a surface. (The surface stress σ^{ESB} results from the totality of the cohesive forces in the near-surface layer). The cohesive forces are caused by the electron exchange in the near-surface layer.

Remark

For an offspring surface state So, we have to use instead of the Eq. (5.26) the Eq. (5. 26a)

$$
n(E)dE \rightarrow \left(n'(E_k) + \delta(E_k - E^{So}) \right) n(E_i) n(E_j) dE_k dE_i dE_j. \tag{5.26a}
$$

The appearance of an offspring surface state is accompanied by a reduction of the state numbers in the bulk.

The Eq.

$$
\int_{E^B}^{E^T} n'(E)dE = \int_{E^B}^{E^T} n(E)dE - 1. \tag{5.36}
$$

has an infinite number of solutions as, e.g., the Eqs. (7.6)–(7.8)

(1) $$n'(E) = n(E) - \delta(E - E^C) \quad \text{with } E^B < E^C < E^T,$$ (7.6)

(2) $$n'(E) = n(E) - \frac{1}{E^T - E^B} \quad \text{and}$$ (7.7)

(3) $$n'(E) = \left(1 - \frac{1}{N}\right)n(E).$$ (7.8)

7.2 The Point of Zero Charge (PZC) and the Fermi Level Shift

We consider in a semi-infinitely extended three-dimensional body the charge density of the electrons in the surface band q^{ESB} belonging to the surface state E_k^S

$$q^{\text{ESB}} = \frac{-e}{L_i L_j} \int_0^\infty p(E_k^S + E_{i+j} - \zeta_k - \zeta_i - \zeta_j) n(E_{i+j}) dE_{i+j}.$$ (7.9)

(E_k^S may be the energy of the additional surface state Sa or the offspring surface state So.)
For

$$\zeta_k = E_k^S$$ (7.10)

the right-hand side of Eq. (7.9) takes the form

$$\frac{-e}{L_i L_j} \int_0^\infty p(E_{i+j} - \zeta_i - \zeta_j) n(E_{i+j}) dE_{i+j}.$$ (7.11)

The term in Eq. (7.11) can also be interpreted as the electron charge density of an arbitrary plane in the body which is oriented parallel to the x_i- and x_j-axes.

As already mentioned in Sect. 2.2, Kronig and Penney have not given any criterion for the position of the Fermi level in their model. In the thermal equilibrium and for a given position of the Fermi level, the charge of the electrons in the mentioned fictive plane is compensated by the charge of the atomic kernels and the *total* charge density of this plane is zero.

Therefore, the *total* surface charge density is given by the difference between q^{ESB} in Eq. (7.9) and the term in Eq. (7.11).

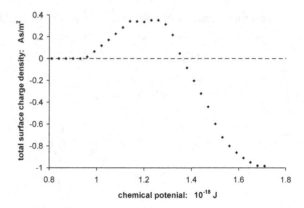

Fig. 7.3 The dependence of the *total* surface charge density q^{tot} at $T = 300$ K on the chemical potential ζ calculated with the lower and the upper surface states for a semi-infinite body

If we consider more than one surface state, we have to use the Eq. (7.12) for the calculation of the *total* surface charge density q^{tot}

$$q^{tot} = \sum_{v} \left\{ \frac{-e}{L_i L_j} \int_0^\infty p(E_{kv}^S + E_{i+j} - \zeta_k - \zeta_i - \zeta_j)n(E_{i+j})dE_{i+j} \right.$$

$$\left. - \frac{-e}{L_i L_j} \int_0^\infty p(E_{i+j} - \zeta_i - \zeta_j)n(E_{i+j})dE_{i+j} \right\}. \tag{7.12}$$

Taking into account the partially occupied lower and the upper surface states, we obtain for the dependence of the *total* surface charge density on the chemical potential ζ the graph in Fig. 7.3.

The point of zero charge (PZC) is a *pH* value. For a solid in contact with the vacuum, the chemical potential at the PZC amounts to $\zeta^{PCZ} = 1.36 \times 10^{-18}$ J.

7.3 The Contribution of the Electrostatic Repulsion Between the Electrons in the Surface Bands to the Surface Energy

The results of the calculations of the surface energy of the electrons in the surface bands φ^{ESB} and of the surface free energy carried out in foregoing section are only correct in the PZC.

If an excess charge is located in the surface layer, we have to take into account the electrostatic repulsion between the surface charges.

The differential of the electrostatic potential V of an electron (or a hole) in a charged plane is

$$dV = \frac{eq}{4\pi\varepsilon_0\varepsilon r}dA. \qquad (7.13)$$

The symbols have the following meanings:
e elementary electric charge,
q charge density
dA differential of area of the surface plane,
ε dielectric constant, and
ε_0 permittivity

For the electrostatic potential V of an electron (or a hole) in the center of a circular disc, we obtain

$$V = \int_{R_0}^{R} \frac{eq^{ESB}}{4\pi\varepsilon_0\varepsilon r}2\pi r dr = \frac{e^2 Z(\zeta)}{2\varepsilon_0\varepsilon\pi R^2}\int_{R_0}^{R} dr = \frac{e^2 Z(\zeta)}{2\varepsilon_0\varepsilon\pi R^2}(R - R_0). \qquad (7.14)$$

Here, R_0 is the radius of an area free of charges surrounding a single electron (or hole) under consideration and $Z(\zeta)$ means here the number of elementary charges.

Neglecting the effects of the boundaries, we approximate the potential electro-static energy of all elementary charges in the surface layer per unit of the surface area by the Eq. (7.15)

$$\varphi^{elstat} \approx \frac{Z(\zeta)V}{\pi R^2} = \frac{e^2 Z^2(\zeta)}{2\varepsilon_0\varepsilon\pi^2 R^4}(R - R_0). \qquad (7.15)$$

Using the relations

$$\pi R^2 = L_i L_j \quad \text{and} \qquad (7.16)$$

$$\pi R_0^2 = \frac{L_i L_j}{Z(\zeta)} \qquad (7.17)$$

we obtain for the potential electrostatic energy per unit of surface of all elementary charges in the surface layer

$$\varphi^{elstat} \approx \frac{e^2}{2\varepsilon_0\varepsilon\pi^{\frac{1}{2}}(L_i L_j)^{\frac{3}{2}}}\left(Z^2(\zeta) - \frac{Z^2(\zeta)}{\sqrt{|Z(\zeta)|}}\right). \qquad (7.18)$$

According to Eq. (7.12), we have for $Z(\zeta)$

$$Z(\zeta) = \sum_{v} \left\{ \int_{0}^{\infty} p(E_{kv}^{S} + E_{i+j} - \zeta_{k} - \zeta_{i} - \zeta_{j})n(E_{i+j})dE_{i+j} \right.$$
$$\left. - \int_{0}^{\infty} p(E_{i+j} - E_{i}^{S} - E_{j}^{S})n(E_{i+j})dE_{i+j} \right\}. \tag{7.19}$$

In the following calculations, for the dielectric constant ε the value 1 has been used in default of a better knowledge.

The Fig. 7.4 shows the electrostatic contributions to the surface energy caused by the charges in the lower energy band belonging to the lower surface state and for the lower energy band belonging to the upper surface state as a function of the chemical potential ζ calculated for a semi-infinite body with $\varepsilon = 1$ at $T = 300$ K.

A minimum value of the electric potential energy appears in Fig. 7.3 at the PZC.

In the further calculations, we have now to substitute φ^{ESB} by $\varphi^{ESB}-\varphi^{elstat}$. Applying the Herring's formula, Eq. (1.3), to the Eq. (7.18), we find for the effect of the electrostatic repulsion to the total surface stress

$$\sigma_{ii}^{elstat} = \varphi^{elstat} + \frac{\partial \phi^{elstat}}{\partial \varepsilon_{ii}} = \varphi^{elstat} + L_{i}\frac{\partial \varphi^{elstat}}{\partial L_{i}}$$
$$= -\frac{1}{2}\frac{e^{2}}{2\varepsilon_{0}\varepsilon\pi^{\frac{1}{2}}(L_{i}L_{j})^{\frac{3}{2}}}\left(Z^{2}(\zeta) - \frac{Z^{2}(\zeta)}{\sqrt{|Z(\zeta)|}} \right). \tag{7.20}$$

For $\varepsilon = 1$, the absolute value of the electrostatic contribution to the surface stress resulting from the charges in the lower energy bands belonging to lower surface

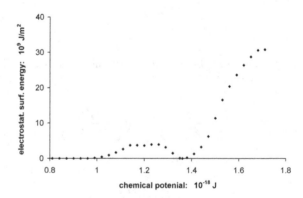

Fig. 7.4 The electrostatic contribution to the surface energy due to the charges in the lower surface energy bands of the lower and the upper surface states versus the chemical potential ζ calculated with $\varepsilon = 1$ for a semi-infinite body at $T = 300$ K

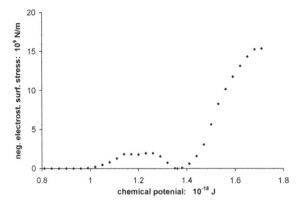

Fig. 7.5 For a semi-infinite body at $T = 300$ K and $\varepsilon = 1$, the absolute value of the electrostatic contribution to the surface stress resulting from the charges in the lower energy bands belonging to the lower and the upper surface states is depicted versus the chemical potential ζ

state and the upper surface state is depicted in Fig. 7.5 as a function of the chemical potential ζ for a semi-infinite body at 300 K.

At the point of zero charge, the surface free energy φ equals the contribution of the electron transition to the surface free energy φ^{TR}.

Remark

For metals, the dielectric function reaches very high values at low frequencies of the electric field. The correct value for zero frequency is not known, see Brauer [1].

All the uncertainties of this last section vanish for $q = 0$.

Reference

1. Brauer W (1972) Einführung in die Elektronentheorie der Metalle, Akademische Verlagsgesellschaft Geest & Portig Leipzig, 85–102

Chapter 8
A Model with a Limited Number of Potential Wells

Abstract As for a semi-infinitely extended body, also for a limited body, principally the same parameters as the surface energy of the electrons in the surface band φ^{ESB} and the surface free energy φ can be calculated. Applying the Herring's Equation to the surface energies the surface stresses σ^{ESB} and σ can be determined. The calculations of the surface charge density q^{ESB} and the point of zero charge PZC have also been done. The main results for the surface energies and the surface stresses calculated for the semi-infinite body and for the nanocube are nearly the same. But, there exist some differences. In the nanocube only the upper surface state appears clearly. The main distinction is a completely different calculation procedure. We combine three one-dimensional bodies with 10 potential wells to a nanocube of $10 \times 10 \times 10$ atoms. This model is more suitable for the discussion of nanoparticles. The calculations of the wave functions of the electrons in the limited body can be executed without problems. The calculation of the surface free energy φ discloses a new aspect.

Keyword Surface of nanocube

8.1 Modeling a Nanoparticle and a Solid Surface

As it has been done in the foregoing chapters, also in this one the allowed energy levels were calculated for a Meander-like potential run in the bulk as it is symbolized in Fig. 8.1. Again, the widths of the potential wells a and of the potential barriers b in the bulk are $a = b = 2 \times 10^{-10}$ m and we assume as in the foregoing chapters for the heights of the barriers in the bulk $U_i = 0.8 \times 10^{-18}$ J and of the potential step at the surface $U^S = 1.6 \times 10^{-18}$ J.

© The Author(s) 2015
W. Gräfe, *Quantum Mechanical Models of Metal Surfaces and Nanoparticles*,
SpringerBriefs in Applied Sciences and Technology,
DOI 10.1007/978-3-319-19764-7_8

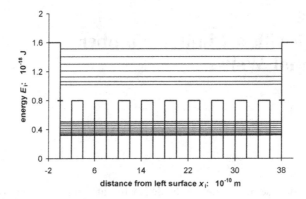

Fig. 8.1 A one-dimensional schematic representation of the potential energies in the bulk and at the surfaces of a body with 10 potential wells; The extended horizontal lines represent the energy levels of the electrons attributed to the "energy bands". The short bars symbolize the surface states; *Source* Prot Met Phys Chem+ in print

8.2 The Energy of the Electrons in the Bulk and in the Surface Bands

Now, we consider three one−dimensional models with only ten potential wells.

Figures 8.1 and 8.2 show the ten energy levels in the "lower energy band". The two short bars at the left and right side symbolize additional surface states with the energy E^S (surface).

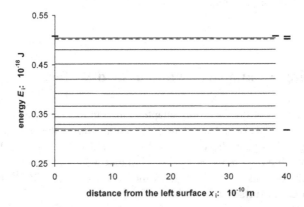

Fig. 8.2 The 10 energy levels for the "lower energy band" calculated for the model depicted in Fig. 8.1; the *continuous lines* are the energy eigenvalues. The *dashed lines* mark the boundaries of the lower allowed energy band for an infinitely extended body calculated with the formulae given by Kronig and Penney, see Fig. 2.3. The rightmost bars are the lowest and the both highest energy levels calculated for a one-dimensional model with 100 potential wells. They can be assigned to the "lower energy band"

In Fig. 8.2 the lowest dashed line symbolizes the bottom of the lower allowed energy band E^B (bottom) for the infinitely extended model of Kronig and Penney, see also Fig. 2.3. The higher dashed line is the top of the lower allowed energy band E^T (top).

The rightmost bars are the lowest and the both highest energy levels calculated for a linear model with 100 potential wells. They can be assigned to the lower energy band.

In Fig. 8.1 the highest energy levels comply with an upper band of allowed energy states. (Three further electronic states corresponding to the "upper band" cannot exist in our model because the energy barriers at the surfaces are to low.)

In the case of a three-dimensional model with a separable potential the discrete energy levels corresponding to the allowed energy bands are sums consisting of the three summands $E_k + E_i + E_j$. Each of the summands belongs to the motion of the electrons in one of the directions x_k, x_i or x_j. Such an energy level belongs to the wave function $\psi_k \times \psi_i \times \psi_j$ in the bulk of the solid.

For the total energy of the electrons in the bulk of the three-dimensional system E^{bulk} we have

$$E^{\text{bulk}} = \sum_{\mu} (E_k + E_i + E_j)_\mu n_\mu^{(3)} \frac{1}{1 + e^{\frac{(E_k+E_i+E_j)_\mu - \zeta}{kT}}} \tag{8.1}$$

with the chemical potential ζ of the three-dimensional system. The sub-index μ is the number of the positions of the sums $(E_k + E_i + E_j)$ in their linear alignment according to their magnitude. The quantity $n_\mu^{(3)}$ means the degeneration of the energy level $(E_k + E_i + E_j)_\mu$. (The sequencing of the energy sums according to their

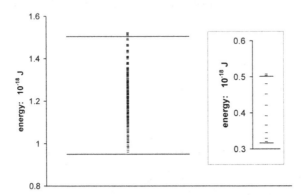

Fig. 8.3 Energy levels in the bulk of a three-dimensional model and of a one-dimensional one; the *longer horizontal lines* mark the boundaries of the allowed energy bands for infinitely extended bodies. The *bars* illustrate the energy levels for models with a limited number of potential wells. On the *left*, the energy band for an infinite three-dimensional model and the energy levels corresponding to a model with 10 × 10 × 10 potential wells are depicted. The insertion on the *right-hand side* shows the band boundaries and the energy levels for a one-dimensional infinitely extended model and another one with ten potential wells

magnitude with the sub-index μ is not necessary but, it facilitates the survey and the practical calculation for $T = 0$ K.)

In the left part of Fig. 8.3 the energy levels are depicted which result by a combination of three one-dimensional models to a three-dimensional one.

8.3 Calculation of the Surface Parameters

In Fig. 8.4 the graphs of the probability densities of the electrons are depicted for three eigenvalues. The lowest curve is the probability density for the electron with the lowest eigenvalue in the "lower energy band". The midway graph belongs to the eigenvalue directly below the highest eigenvalue and reveals a concentration of the probability density of this electron directly at the surface as we would expect it for a surface state. The highest curve is the probability density for the highest eigenvalue and resembles a surface resonance.

Figure 8.5 shows the sum of all probability densities of the electrons in the "lower energy band"

In Fig. 8.5 we find a near-surface enhancement of the sum of all probability densities for the electrons in the "lower energy band". By this, there is also an enhancement of the binding forces parallel to the surface and in the immediate vicinity of the surface. In conformity with the statements in Chaps. 4 and 14 and in [1] this explains the existence of a fatigue limit which is greater than zero.

The electrons located at the surface perpendicular to the x_k–axis have the energies $E_k^S + E_i + E_j$. Herein E_k^S is the energy of a surface state in the one-dimensional sub-system belonging to x_k. The wave function of an electron in a surface state is $\psi_k^S \psi_i \psi_j$. (For a surface state at an edge we have for instance the

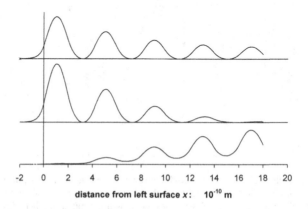

Fig. 8.4 Graphs of the probability densities of the electron with the lowest energy in the "lower energy band" (*lowest curve*), for the electron with the highest energy in the "lower energy band" (*topmost curve*) and for the electron with an energy directly below the highest eigenvalue belonging to the "lower energy band" (*midway curve*); *Source* Prot Met Phys Chem+ in print

Fig. 8.5 The sum of the probability densities of all electrons in the "lower energy band"; *Source* Prot Met Phys Chem+ in print

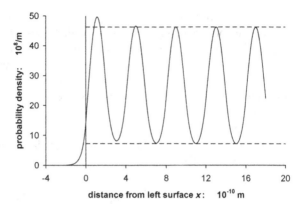

energy value $E_k^S + E_i^S + E_j$ and the energy for an electron at a cubic summit is $E_k^S + E_i^S + E_j^S$.)

8.3.1 The Surface Energy φ^{ESB} of the Electrons in a Surface Band of a Nanocube

Now we consider only the energies of the electrons located at the surface of a nanocube.

Depicted by bars, the Fig. 8.6 shows the discrete energy levels $E_k^S + E_i + E_j$ for the electrons localized in the two-dimensional "surface energy band" of a nanocube

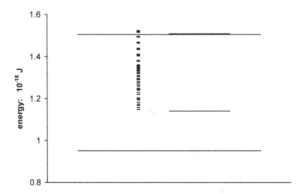

Fig. 8.6 Electron energy levels at a surface of a nanocube with $10 \times 10 \times 10$ potential wells; the *bars* are the energy levels for electrons in the near-surface layer for a three-dimensional model with $10 \times 10 \times 10$ potential wells. On the *right hand side* the *shorter lines* represent the boundaries of the energy band for the electrons localized at the surface of a semi-infinitely extended three-dimensional body. The *long lines* are the boundaries of the allowed energy band in the bulk of an infinitely extended three-dimensional body

with $10 \times 10 \times 10$ potential wells. The boundaries of the allowed energy band in the bulk of the corresponding infinitely extended body are symbolized by long lines. Theses boundaries have been calculated with the approximation of the three-dimensional solid by three infinitely extended one-dimensional models. The boundaries of the surface energy band for the semi-infinitely extended three-dimensional model are designed by the shorter lines in the right part of the Fig. 8.6.

For the nanocube we can calculate the energy density of the electrons at one of the surfaces φ^{ESB} for an arbitrary position of the chemical potential and at any temperature using the formula

$$\varphi^{ESB} = \frac{1}{2} \frac{1}{L_i L_j} \sum_{\mu'} (E_k^S + E_i + E_j)_{\mu'} n_{\mu'}^{(2)} \frac{1}{1 + e^{\frac{(E_k^S + E_i + E_j)_{\mu'} - \zeta}{kT}}}. \tag{8.2}$$

For the two-dimensional manifold of the energy levels in a "surface band" the degeneration $n^{(2)}$ and the sub-index μ' are used.

Figure 8.7 shows the quantity φ^{ESB} for the lower "surface band" of a nanocube with $10 \times 10 \times 10$ potential wells versus the chemical potential for the three-dimensional body $\zeta = \zeta_k + \zeta_i + \zeta_j$ at the temperature $T = 0$ K.

8.3.2 Calculation of the Surface Free Energy φ in a Nanocube

Due to the generation of a potential step at the surface the energy eigenvalues of the electrons in the solid increase. For the body under consideration the surface free

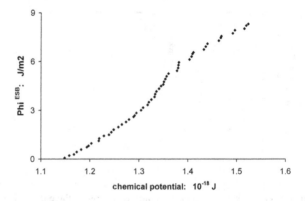

Fig. 8.7 The surface energy for the electrons in the lower "surface band" φ^{ESB} at $T = 0$ K for a nanocube with $10 \times 10 \times 10$ potential wells depicted versus the chemical potential for the three-dimensional body $\zeta = \zeta_k + \zeta_i + \zeta_j$; *Source* Prot Met Phys Chem+ in print

energy φ is the sum of the increments of the energy eigenvalues per total surface area. For the practical calculation of the surface free energy of a nanocube φ we calculate the difference between the total energy of a nanocube with surfaces and the total energy of the same cube without any surface.

$$
6\varphi = \frac{1}{L_iL_j}\left\{\sum_{\mu=1}^{M+6}(E_k+E_i+E_j)_\mu n_\mu^{(3)}\frac{1}{1+e^{\frac{(E_k+E_i+E_j)_\mu-\zeta}{kT}}}\right\}
$$
$$
-\frac{1}{L_iL_j}\left\{\sum_{\mu=1}^{M}(E_k^{ul}+E_i^{ul}+E_j^{ul})_\mu n_\mu^{(3)}\frac{1}{1+e^{\frac{(E_k^{ul}+E_i^{ul}+E_j^{ul})_\mu-\zeta}{kT}}}\right\}. \tag{8.3}
$$

The symbol E^{ul} means an eigenvalue for an unlimited system consisting of a periodical repetition of L/c energy wells. For the calculation of the energy levels E^{ul} we invert the procedure described in the reference to Figs. 2.1 and 2.2. However, here we have to start with a given value of the wave number k.

The consideration of the surface states E^S at the six surfaces has been symbolized by the upper limit of summation $M + 6$.

From the first term on the right hand side in Eq. (8.3) we can separate the quantity

$$
6\varphi^{ESB} = 6\frac{1}{L_iL_j}\sum_{\mu'}(E_k^S+E_i+E_j)_{\mu'}n_{\mu'}^{(2)}\frac{1}{1+e^{\frac{(E_k^S+E_i+E_j)_{\mu'}-\zeta}{kT}}}. \tag{8.4}
$$

So we obtain for the surface free energy of the nanocube φ

$$
\varphi = \varphi_k^{ESB}
$$
$$
+\frac{1}{6L_iL_j}\left\{\sum_\mu^{M}(E_k+E_i+E_j)_\mu n_\mu^{(3)}\frac{1}{1+e^{\frac{(E_k+E_i+E_j)_\mu-\zeta}{kT}}}\right\}
$$
$$
-\frac{1}{6L_iL_j}\left\{\sum_\mu^{M}(E_k^{ul}+E_i^{ul}+E_j^{ul})_\mu n_\mu^{(3)}\frac{1}{1+e^{\frac{(E_k^{ul}+E_i^{ul}+E_j^{ul})_\mu-\zeta}{kT}}}\right\}. \tag{8.5}
$$

For the nanocube with six free surfaces consisting of $10 \times 10 \times 10$ atoms, Fig. 8.8 shows the surface free energy calculated for $T = 0$ K versus the chemical potential ζ. These calculations were carried out under condition that the additional surface states E^S are empty, that means the surface energy φ^{ESB} is zero.

The contribution to the surface free energy of the surface state at $E^S = 0.8 \times 10^{-18}$ J and the lower "energy band" (for the nanocube consisting of $10 \times 10 \times 10$ atoms and with six free surfaces) at $T = 0$ K is depicted in Fig. 8.9 versus the chemical potential of the three-dimensional body $\zeta = \zeta_k + \zeta_i + \zeta_j$.

In this chapter the electrostatic repulsion between the electrons in the "surface bands" is not taken into consideration for the calculation of the surface free energy.

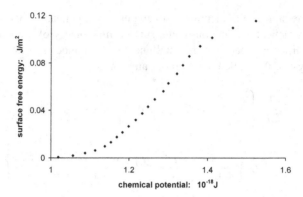

Fig. 8.8 Surface free energy φ of a nanocube with six free surfaces consisting of $10 \times 10 \times 10$ atoms at $T = 0$ K calculated without regard to the additional surface states E^S and depicted as a function of the chemical potential ζ; *Source* Prot Met Phys Chem+ in print

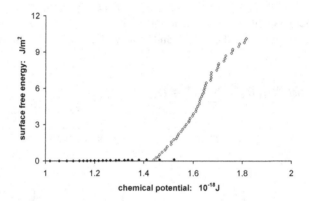

Fig. 8.9 The contribution to the surface free energy φ of the lower "energy band" belonging to the surface state at $E^S = 0.8 \times 10^{-18}$ J (*empty rhombs*) for a nanocube (which consists of $10 \times 10 \times 10$ atoms) with six free surfaces at $T = 0$ K is depicted as a function of the chemical potential ζ. The contributions to the surface free energy without regard to the surface state depicted in Fig. 8.8 are presented by the *full symbols*

8.3.3 Calculation of the Surface Stress for a Nanocube and a Plate-like Body

For the derivation of the surface stress by use of the Herring's equation we need a potential energy. Between the energies discussed in relation to Eq. (6.6), only the eigenvalues $E^e(E, \varepsilon)$ are functions of ε_{ii}. In contrast to the eigenvalues which exist only in the allowed energy bands or discrete states, the variable E appearing in the Fermi-distribution function is defined in the energy range from $-\infty$ to $+\infty$.

As the argument of the Fermi distribution function, the quantity E has not necessarily to have the feature of a special form of energy. Therefore, we consider the argument of the Fermi-distribution E as an independent variable.

According to Herring's Eq. (1.3) we have to perform a partial derivation with respect to the strain. But, the Fermi-distribution does not explicitly depend on the strain. That is, the Fermi-function has not the form $p(E, \varepsilon)$. Consequently, the partial derivation of the Fermi-distribution with respect to the strain ε_{ii} is zero. This question will be discussed in Chap. 13 in more detail.

Degenerated states have the same energy but, they cause different contributions to the surface stress in a given direction. Therefore, we have to consider in the formula for the surface stress each degenerated state separately. For this, in the following we label each individual state by the sub-index μ.

For the calculation of the surface stress we restrict ourselves again to the contributions of the "lower energy band."

We apply Herring's Eq. (1.3) to the surface free energy as it is provided by the Eq. (8.5) and obtain for the surface stress in a nanocube at an arbitrary temperature T and for any position of the chemical potential ζ

$$
\begin{aligned}
\sigma_{ii} &= \varphi \delta_{ii} + \frac{\partial \varphi}{\partial \varepsilon_{ii}} \\
&= \sigma^{\mathrm{ESB}} + \frac{1}{6L_i L_j} \left\{ \sum_{\mu''}^{M''} \left(\frac{\partial E_i}{\partial \varepsilon_{ii}} \right)_{\mu''} \frac{1}{1 + e^{\frac{(E_k + E_i + E_j)_{\mu''} - \zeta}{kT}}} \right\} \\
&\quad - \frac{1}{6L_i L_j} \left\{ \sum_{\mu''}^{M''} \left(\frac{\partial E_i^{\mathrm{ul}}}{\partial \varepsilon_{ii}} \right)_{\mu''} \frac{1}{1 + e^{\frac{(E_k^{\mathrm{ul}} + E_i^{\mathrm{ul}} + E_j^{\mathrm{ul}})_{\mu''} - \zeta}{kT}}} \right\}.
\end{aligned}
\tag{8.6}
$$

In Eq. (8.6) σ^{ESB} means

$$
\sigma^{\mathrm{ESB}} = \frac{1}{L_i L_j} \left\{ \sum_{\mu''}^{M''} \left(\frac{\partial E_i}{\partial \varepsilon_{ii}} \right)_{\mu''} \frac{1}{1 + e^{\frac{(E_k^S + E_i + E_j)_{\mu''} - \zeta}{kT}}} \right\}.
\tag{8.7}
$$

Let the surface state E^S be completely empty at the temperature $T = 0$ K but, all other energy levels completely occupied. Then from Eq. (8.6) follows

$$
\sigma_{ii} = -\frac{1}{6} \frac{1}{L_i L_j} \left\{ \sum_{\mu''}^{M''(\zeta)} \left(\frac{\partial E_i}{\partial b_i} \right)_{\mu''} - \sum_{\mu''}^{M''(\zeta)} \left(\frac{\partial E_i^{\mathrm{ul}}}{\partial b_i} \right)_{\mu''} \right\}.
\tag{8.8}
$$

For more clearness we repeat the foregoing calculations for a plate-like body. That is, we consider a limited body which is infinitely extended into the x_i- and the x_j-directions but, limited by two surfaces perpendicular to the x_k-direction. The distance between the both surfaces is L_k. Let this plate-like body be subdivided into identical sections with the sizes L_i and L_j. We compare this limited body with another hypothetical one which is infinitely extended into the three dimensions but, subdivided into identical sections with the sizes L_k, L_i, and L_j.

Here too, we consider the difference between the sums of the electron energies in the corresponding sections with the dimensions L_k, L_i, and L_j in the both hypothetical bodies and divide the result by the area $L_i \times L_j$. In this way we obtain for the double value of the surface free energy for the plate-like body φ^{pl}

$$2\varphi^{\text{pl}} = \frac{1}{L_i L_j} \left\{ \sum_{\mu''=1}^{M''+2} (E_k + E_i^{\text{ul}} + E_j^{\text{ul}})_{\mu''} \frac{1}{1 + e^{\frac{(E_k + E_i^{\text{ul}} + E_j^{\text{ul}})_{\mu''} - \zeta}{kT}}} \right\}$$
$$- \frac{1}{L_i L_j} \left\{ \sum_{\mu''=1}^{M''} (E_k^{\text{ul}} + E_i^{\text{ul}} + E_j^{\text{ul}})_{\mu''} \frac{1}{1 + e^{\frac{(E_k^{\text{ul}} + E_i^{\text{ul}} + E_j^{\text{ul}})_{\mu''} - \zeta}{kT}}} \right\}. \tag{8.3a}$$

The consideration of the states E^{S} at the two surfaces is here symbolized by the upper limit of summation $M'' + 2$.

From the first term on the right hand side in Eq. (8.3a) we can separate the quantity

$$2\varphi^{\text{ESB,pl}} = 2 \frac{1}{L_i L_j} \sum_{\mu''} (E_k^{\text{S}} + E_i^{\text{ul}} + E_j^{\text{ul}})_{\mu''} \frac{1}{1 + e^{\frac{(E_k^{\text{S}} + E_i^{\text{ul}} + E_j^{\text{ul}})_{\mu''} - \zeta}{kT}}}. \tag{8.2a}$$

So we obtain for the surface free energy φ^{pl} of the plate-like body

$$\varphi^{\text{pl}} = \varphi^{\text{ESB,pl}}$$
$$+ \frac{1}{2L_i L_j} \left\{ \sum_{\mu''}^{M''} (E_k + E_i^{\text{ul}} + E_j^{\text{ul}})_{\mu''} \frac{1}{1 + e^{\frac{(E_k + E_i^{\text{ul}} + E_j^{\text{ul}})_{\mu''} - \zeta}{kT}}} \right\}$$
$$- \frac{1}{2L_i L_j} \left\{ \sum_{\mu''}^{M''} (E_k^{\text{ul}} + E_i^{\text{ul}} + E_j^{\text{ul}})_{\mu''} \frac{1}{1 + e^{\frac{(E_k^{\text{ul}} + E_i^{\text{ul}} + E_j^{\text{ul}})_{\mu''} - \zeta}{kT}}} \right\}. \tag{8.5a}$$

If we apply Herring's Equation (1.3) to the surface free energy as it is provided by the Eq. (8.5a), we obtain the surface stress in the plate-like body at an arbitrary temperature T and for any position of the chemical potential ζ

$$\sigma_{ii}^{\mathrm{pl}} = \varphi^{\mathrm{pl}} \delta_{ii} + \frac{\partial \varphi^{\mathrm{pl}}}{\partial \varepsilon_{ii}}$$

$$= \sigma^{\mathrm{ESB,pl}} + \frac{1}{2L_i L_j} \left\{ \sum_{\mu''}^{M''} \left(\frac{\partial E_i^{\mathrm{ul}}}{\partial \varepsilon_{ii}} \right)_{\mu''} \frac{1}{1 + e^{\frac{(E_k + E_i^{\mathrm{ul}} + E_j^{\mathrm{ul}})_{\mu''} - \zeta}{kT}}} \right\} \tag{8.6a}$$

$$- \frac{1}{2L_i L_j} \left\{ \sum_{\mu''}^{M''} \left(\frac{\partial E_i^{\mathrm{ul}}}{\partial \varepsilon_{ii}} \right)_{\mu''} \frac{1}{1 + e^{\frac{(E_k^{\mathrm{ul}} + E_i^{\mathrm{ul}} + E_j^{\mathrm{ul}})_{\mu''} - \zeta}{kT}}} \right\}.$$

In Eq. (8.6a) $\sigma^{\mathrm{ESB,pl}}$ means

$$\sigma^{\mathrm{ESB,pl}} = \frac{1}{2} \frac{1}{L_i L_j} \left\{ \sum_{\mu''=1}^{M''} \left(\frac{\partial E_i^{\mathrm{ul}}}{\partial \varepsilon_{ii}} \right)_{\mu''} \frac{1}{1 + e^{\frac{(E_k^S + E_i^{\mathrm{ul}} + E_j^{\mathrm{ul}})_{\mu''} - \zeta}{kT}}} \right\}. \tag{8.7a}$$

Here too, we assume that the surface states E^S are completely empty at the temperature $T = 0$ K but, all other energy levels are completely occupied. Then we obtain instead of Eq. (8.8)

$$\sigma_{ii}^{\mathrm{pl}} = -\frac{1}{2} \frac{1}{L_i L_j} \left\{ \sum_{\mu''}^{M''(\zeta)} \left(\frac{\partial E_i^{\mathrm{ul}}}{\partial b_i} \right)_{\mu''} - \sum_{\mu''}^{M''(\zeta)} \left(\frac{\partial E_i^{\mathrm{ul}}}{\partial b_i} \right)_{\mu''} \right\} = 0. \tag{8.8a}$$

In this special case, it vanishes the sum of all stresses in the total system consisting in the surface layer and the bulk.

The derivations of $\varphi^{\mathrm{ESB,pl}}$ result in the forces in the near-surface layer. In contrast to this, in the plate-like model described here, the derivations of the further eigenvalues E_i^{ul} remaining in Eq. (8.6a) yield the variations of stresses in the bulk.

Now we return to the nanocube with $10 \times 10 \times 10$ potential wells and calculate the surface stress σ^{ESB} at $T = 0$ K caused by the electrons in the "lower energy band" belonging to the additional surface state E^S.

If the solid is stretched in the x_i-direction, we obtain for $T = 0$ K the data depicted versus the chemical potential of the three-dimensional body $\zeta = \zeta_k + \zeta_i + \zeta_j$ in Fig. 8.10. For comparison, the surface stress values at 300 K for the semi-infinitely extended body calculated with the convolution integral in Chap. 6 are also presented in this picture.

For the numerical calculations of $\sigma_{ii}^{\mathrm{ESB}}$ the approximation

$$\frac{\partial (E_i)_{\mu''}}{\partial b_i} \approx \frac{\Delta (E_i)_{\mu''}}{\Delta b_i}. \tag{8.9}$$

has been used with the quantity $\Delta b/b$ amounting to 1 %. The surface stress has also been calculated with the value $\Delta b/b = 0.5\,\%$. The data points calculated with the values of $\Delta b/b = 0.5\,\%$ and 1 % coincide in the scale of Fig. 8.10.

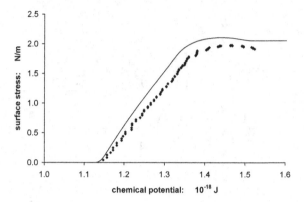

Fig. 8.10 Contribution to the surface stress σ^{ESB} generated by the electrons in the "lower energy band" belonging to the additional surface state E^S; The data were calculated for a nanocube with $10 \times 10 \times 10$ potential wells at $T = 0$ K. These results are depicted by *rhombs* as a function of the chemical potential of the three-dimensional body $\zeta = \zeta_k + \zeta_i + \zeta_j$. The *curve* represents the values of the surface stress at 300 K which were calculated for the same lattice parameters with the convolution integral shown in Fig. 6.2

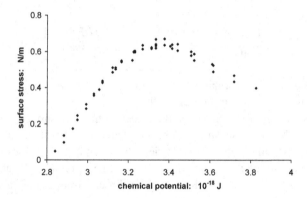

Fig. 8.11 Contribution to the surface stress σ^{ESB} generated by the electrons in the "upper energy band" of the additional surface state E^S for a nanocube with $10 \times 10 \times 10$ potential wells at $T = 0$ K versus the chemical potential of the three-dimensional body $\zeta = \zeta_k + \zeta_i + \zeta_j$

The surface stress values σ^{ESB} generated by the electrons in the "upper energy band" belonging to the additional surface state E^S are pictured in Fig. 8.11.

8.4 Remarks

(1) The surface energy of the electrons in the surface bands φ^{ESB} differs from the surface free energy φ. The quantity φ^{ESB} characterizes an existing surface,

whereas the surface free energy φ has to be transferred from the outside into the solid for the generation of a new surface. The surface free energy φ is a thermodynamic quantity of the total system consisting of the volume and the surface.

(2) As we can see it from Fig. 8.5 the probability density of the electrons is dislocated from the bulk to the surface. If due to the generation of the surface the number of electrons remains unchanged in the solid, Eq. (8.8a) means that the stress enhancement at the surface is accompanied by a stress reduction in the bulk. In comparison with φ^{ESB} und σ^{ESB} the surface free energy φ and the surface stress σ calculated with φ are reduced by contributions having already existed in the bulk before the generation of the surface, and being assigned to the surface in φ^{ESB} und σ^{ESB}.

(3) As it is explained in more detail in Sect. 14.4, for a discussion of the origin of the fatigue limit the gradient in the x_k-direction of the surface stress is of interest. This gradient is determined by the wave functions $\psi(x_k)$. The grad σ^{ESB} describes the forces acting on a surface atom whereas the grad σ describes the change of the forces due to the generation of the surface. Therefore, the decisive quantity for the determination of strength and fatigue limit is the grad σ^{ESB}.

8.5 The Surface Charge Densities and the Point of Zero Charge in a Nanocube

In the limited body only the (upper) additional surface state effectively exists.

However, for the sake of illustration we regard the highest energy level of the lower "energy band" in the bulk as the (lower) offspring surface state and obtain for the charge density of the electrons in "surface bands" q^{ESB}

$$q^{ESB} = \frac{-e}{L_i L_j}\left\{\sum_{\mu'}^{M'} n_{\mu'}^{(2)}(E_k^S + E_i + E_j)\frac{1}{1 + e^{\frac{(E_k^S + E_i + E_j)_{\mu'} - \zeta_k - \zeta_i - \zeta_j}{kT}}}\right\}. \tag{8.10}$$

(In Eq. (8.10) the factor ½ as in Eq. (8.5a) does not appear because we consider only one surface.)

The charge density of the electrons in the "surface band" belonging to the "offspring surface state" at T = 0 K for a nanocube with 10 × 10 × 10 potential wells is depicted in Fig. 8.12 as a function of the chemical potential $\zeta = \zeta_k + \zeta_i + \zeta_j$.

Now, we consider an infinitely extended body which is composed by a periodical continuation in three dimensions of a section with $N_k \times N_i \times N_j$ potential wells.

In the thermal equilibrium, the total charge density of a plane oriented parallel to the x_i- and x_j-axes is zero. The density of electrons in this plane is

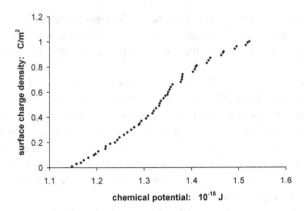

Fig. 8.12 For a nanocube with $10 \times 10 \times 10$ potential wells the charge density of the electrons in the "surface band" belonging to the "offspring surface state" q^{ESB} at $T = 0$ K is depicted versus the chemical potential for the three-dimensional body $\zeta = \zeta_k + \zeta_i + \zeta_j$

$$d_e = \frac{1}{L_i L_j} \left\{ \sum_{\mu'} n_{\mu'}^{(2)} (E_i^{\text{ul}} + E_j^{\text{ul}}) \frac{1}{1 + e^{\frac{(E_i^{\text{ul}} + E_j^{\text{ul}})_{\mu'} - \zeta_i - \zeta_j}{kT}}} \right\}. \tag{8.11}$$

For a plate-like body the *total* surface charge density is determined by the difference between the electron charge density at the surface and the charge density of electrons on a hypothetical plane in the bulk $-ed_e$. Taking into account only the lower "surface band" belonging to the additional (upper) surface state we obtain

$$q^{\text{pl}} = \frac{-e}{L_i L_j} \left\{ \sum_{\mu'}^{M'} n_{\mu'}^{(2)} (E_k^{\text{S}} + E_i^{\text{ul}} + E_j^{\text{ul}}) \frac{1}{1 + e^{\frac{(E_k^{\text{S}} + E_i^{\text{ul}} + E_j^{\text{ul}})_{\mu'} - \zeta_k - \zeta_i - \zeta_j}{kT}}} \right\}$$

$$+ \frac{e}{L_i L_j} \left\{ \sum_{\mu'}^{M'} n_{\mu'}^{(2)} (E_i^{\text{ul}} + E_j^{\text{ul}}) \frac{1}{1 + e^{\frac{(E_i^{\text{ul}} + E_j^{\text{ul}})_{\mu'} - \zeta_i - \zeta_j}{kT}}} \right\}. \tag{8.12}$$

Let us consider the case that the surface states E_k^{S} are completely empty. Due to the generation of the two surfaces perpendicular to the x_k-axes in the plate-like body all energy levels E_k^{ul} are shifted upwards to E_k. Due to this shift the energy levels E_k are recharged.

As it is illustrated in Fig. 8.5, in the one-dimensional sub-system of the electrons moving normally to the surface the total probability density of the electrons in the lower "energy band" is enhanced near the surfaces. That means, it occurs a surface charge without the clear appearance of the (lower) offspring surface state. For the electrons moving into the direction x_k normal to the surface we can formally divide the total probability density into a part representing a constant electron distribution in the bulk and an additional probability density evanescing into the direction away

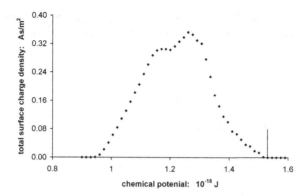

Fig. 8.13 The total surface charge density q^{pl} at $T = 300$ K versus the chemical potential ζ calculated for the lower surface "energy band" in a section of the plate-like body with $10 \times 10 \times 10$ potential wells; The calculations have been carried out for a supposed normalized probability density of the additional electrons near the surface. The short perpendicular line at the right marks the point of zero charge at $\zeta^{PZC} = 1.529 \times 10^{-18}$ J

from the surface. The contribution of the decaying electron density corresponds to the offspring surface state So in the semi-infinite model.

For an approximate determination of the total surface charge density of the plate-like body we handle again the highest energy level $\hat{E}_k$ in the lower "energy band" belonging to the sub-system for x_k as a surface state. In doing so, we obtain the equation

$$q^{pl} \propto \frac{-e}{L_i L_j} \left\{ \sum_{\mu'}^{M'} n_{\mu'}^{(2)} (\hat{E}_k + E_i^{ul} + E_j^{ul}) \frac{1}{1 + e^{\frac{(\hat{E}_k + E_i^{ul} + E_j^{ul})_{\mu'} - \zeta_k - \zeta_i - \zeta_j}{kT}}} \right.$$
$$\left. - \sum_{\mu'}^{M'} n_{\mu'}^{(2)} (E_i^{ul} + E_j^{ul}) \frac{1}{1 + e^{\frac{(E_i^{ul} + E_j^{ul})_{\mu'} - \zeta_i - \zeta_j}{kT}}} \right\}. \tag{8.13}$$

In Eq. (8.13) the symbol for the proportionality $\propto$ has been used because the amount of the enhanced electron density in Fig. 8.5 has not been determined.

For the section of the plate-like body with two free surfaces containing $10 \times 10 \times 10$ potential wells the total surface charge density has been calculated with the Eq. (8.13). The calculation was carried out with a supposed normalized probability density for the additional electrons near the surface. Figure 8.13 shows the results.

The chemical potential at which the total surface charge density is zero amounts to $\zeta^{PZC} = 1.529 \times 10^{-18}$ J. This value agrees at best with the triple energy of the lower surface state (So) in the one-dimensional semi-infinite model, $3 \times E_k^{So} = 1.519 \times 10^{-18}$ J.

If we substitute E for E^{ul} in Eq. (8.12), the results are almost the same. This latter case corresponds to the consideration of a nanocube instead of the plate-like body.

Reference

1. Gräfe W (1989) Cryst Res Technol 24(9):879

Chapter 9
Surface Stress-Charge Coefficient (Estance)

Abstract The surface stress-charge coefficient (Estance) has been calculated for the semi-infinitely extended three-dimensional body and for a nanocube with $10 \times 10 \times 10$ atoms. For the semi-infinite body the calculated quantity $\varsigma_{ii} = d\sigma_{ii}/dq$ amounts to nearly -1.34 V in the range of the chemical potential between 1.1×10^{-18} J and 1.4×10^{-18} J.

Keywords Estance · Surface stress-charge coefficient

From Eq. (6.5) the derivation of the surface stress with respect to the chemical potential is as follows

$$\frac{d\sigma_{ii}^{ESB}}{d\zeta} = \frac{c_i}{L_i L_j} \int_0^\infty \frac{\partial p(E_k^S + E_{i+j} - \zeta)}{\partial \zeta} \int_0^\infty \frac{\partial E_i}{\partial b_i} n(E_i) n(E_{i+j} - E_i) dE_i dE_{i+j}$$

$$+ \frac{c_i}{L_i L_j} \int_0^\infty \frac{\partial p(E_k^S + E_{i+j} - \zeta)}{\partial \zeta} (E_k^S + E_{i+j}) \int_0^\infty \frac{\partial n(E_i)}{\partial b_i} n(E_{i+j} - E_i) dE_i dE_{i+j}.$$

$$(9.1)$$

Correspondingly, with Eq. (6.7) we obtain the derivation of the surface charge density with respect to the chemical potential

$$\frac{dq^{ESB}}{d\zeta} = -\frac{e}{L_i L_j} \int_0^\infty \frac{\partial p(E_k^S + E_{i+j} - \zeta)}{\partial \zeta} \int_0^\infty n(E_i) n(E_{i+j} - E_i) dE_i dE_{i+j}. \qquad (9.2)$$

The contribution of the surface states to the surface stress-charge coefficient (estance) is given by the diagonal tensor components $\varsigma_{ii} = d\sigma_{ii}/dq$

© The Author(s) 2015
W. Gräfe, *Quantum Mechanical Models of Metal Surfaces and Nanoparticles*,
SpringerBriefs in Applied Sciences and Technology,
DOI 10.1007/978-3-319-19764-7_9

$$\varsigma_{ii}(\zeta) = \frac{d\sigma_{ii}}{dq} = \frac{d\sigma_{ii}}{d\zeta}\frac{d\zeta}{dq} =$$

$$= c_i \left\{ \int\limits_0^\infty \frac{\partial p(E_k^S + E_{i+j} - \zeta)}{\partial\zeta} \int\limits_0^\infty \frac{\partial E_i}{\partial b_i} n(E_i)n(E_{i+j} - E_i)dE_idE_{i+j} \right.$$

$$\left. + \int\limits_0^\infty \frac{\partial p(E_k^S + E_{i+j} - \zeta)}{\partial\zeta}(E_k^S + E_{i+j}) \int\limits_0^\infty \frac{\partial n(E_i)}{\partial b_i} n(E_{i+j} - E_i)dE_idE_{i+j} \right\}.$$

$$\times \left\{ -e \int\limits_0^\infty \frac{\partial p(E_k^S + E_{i+j} - \zeta)}{\partial\zeta} \int\limits_0^\infty n(E_i)n(E_{i+j} - E_i)dE_idE_{i+j} \right\}^{-1}$$

$$(9.3)$$

For the semi-infinite body the calculated quantity $\varsigma_{ii} = d\sigma_{ii}/dq$ amounts to nearly -1.34 V in the range of the chemical potential between 1.1×10^{-18} J and 1.4×10^{-18} J. These data are depicted in Fig. 9.1 as empty rhombs.

For the model with a limited number of potential wells we obtain from Eq. (8.7) for the surface stress σ_{ii}^{ESB} and from Eq. (8.10) for the charge density of the electrons in surface band q^{ESB} the following formulae for the estance.

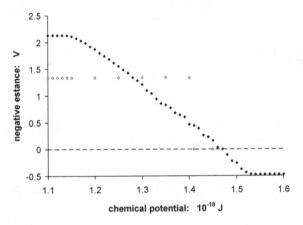

Fig. 9.1 Dependence of the estance $d\sigma_{ii}/dq$ (surface stress-charge coefficient) on the chemical potential ζ; The data for the infinitely extended body are depicted as *empty rhombs* and the results for a nanocube consisting of $10 \times 10 \times 10$ atoms at $T = 300$ K are displayed as *filled symbols*

$$\varsigma_{ii}(\zeta) = \frac{d\sigma_{ii}}{dq} = \frac{d\sigma_{ii}}{d\zeta}\frac{d\zeta}{dq} =$$

$$= \frac{c_i}{2}\sum_{\mu''}^{M''}\frac{\partial(E_i)_{\mu''}}{\partial b_i}\frac{e^{\frac{E_k^S+E_i+E_j-\zeta}{kT}}}{\left(1+e^{\frac{E_k^S+E_i+E_j-\zeta}{kT}}\right)^2}\frac{1}{kT}$$

$$\times\left\{-e\sum_{\mu''}^{M''}\frac{e^{\frac{E_k^S+E_i+E_j-\zeta}{kT}}}{\left(1+e^{\frac{E_k^S+E_i+E_j-\zeta}{kT}}\right)^2}\frac{1}{kT}\right\}^{-1} \qquad (9.4)$$

(Deviating from Eq. (8.10), here we carry out the summations of each degenerated state indexed by μ'' instead of a summation using the degenerations $n_{\mu'}$. See comment to the Eq. (8.6).)

The estance $d\sigma_{ii}/dq$ calculated with Eq. (9.4) shows a distinct dependence on ζ. These results are also presented in Fig. 9.1.

Chapter 10
Regard to the Spin in the Foregoing Texts

Abstract Now we consider the influence of the spin on the foregoing results. For this, we complete the wave functions for the three-dimensional space $\psi(r)$ by the eigenfunctions of the spin operator χ_+ and χ_-

$$\psi(r) \rightarrow \chi_\pm \psi(r).$$

Keyword Spin

Now we consider the influence of the spin on the foregoing results. For this, we complete the wave functions for the three-dimensional space $\psi(r)$ by the eigenfunctions of the spin operator χ_+ and χ_-

$$\psi(r) \rightarrow \chi_\pm \psi(r). \tag{10.1}$$

If the spin is an invariant, this separated ansatz for the wave function is the correct solution of the Schrödinger equation.

Furthermore, we separate the three space coordinates in the wave function and obtain

$$\chi_\pm \psi(x_1)\psi(x_2)\psi(x_3). \tag{10.2}$$

The numerical corrections necessary for the regard to the spin in the foregoing outcomes consist only in a multiplication of the final results for the surface energy φ, the surface stress σ_{ii}, the near-surface stress s_{ii}, and the surface charge density q by the factor 2.

W. Gräfe, *Quantum Mechanical Models of Metal Surfaces and Nanoparticles*,
SpringerBriefs in Applied Sciences and Technology,
DOI 10.1007/978-3-319-19764-7_10

Chapter 11
Detailed Calculation of the Convolution Integrals

Abstract The convolution integrals for calculation of the density of electron states in the surface layer $n_{i+j}(E_{i+j})$ are considered in detail. The numerical values of this density function have been determined. For differentiation of surface energy with respect to the strain ε, according to Herring's equation the extended Leibniz rule has been applied. The area of integration for the practical accomplishment of the convolution procedure is depicted.

Keyword Convolution integral

For the sake of shortness, in the Chaps. 6 and 7 the quantities 0 and $+\infty$ have been used as the bounds of integration in the convolution integrals of the type

$$n_{1+2}(E_{1+2}) = \int_{0}^{+\infty} n_1(E_1)n_2(E_{1+2} - E_1)\mathrm{d}E_1 . \tag{11.1}$$

The correct form of this convolution integrals for integration over the lower allowed energy band from E^B until E^T is

$$n_{1+2}(E_{1+2}) = \int_{E_1^B(\varepsilon)}^{E_1^T(\varepsilon)} n_1(E_1)n_2(E_{1+2} - E_1)\mathrm{d}E_1. \tag{11.2}$$

The graph of the density function $n_{1+2}(E_{1+2})$ resulting from Eq. (11.2) is depicted in Fig. 11.1.

Formally, we have for the surface energy of the electrons in the surface band

$$\varphi^{\mathrm{ESB}} = \frac{1}{L_1 L_2} \int_{E^B(\varepsilon)}^{E^T(\varepsilon)} p\big(E^S + E_{1+2} - \zeta\big)\big(E^S + E_{1+2}\big)n_{1+2}(E_{1+2})\mathrm{d}E_{1+2}. \tag{11.3}$$

© The Author(s) 2015

W. Gräfe, *Quantum Mechanical Models of Metal Surfaces and Nanoparticles*,
SpringerBriefs in Applied Sciences and Technology,
DOI 10.1007/978-3-319-19764-7_11

Fig. 11.1 The graph of the density $n_{1+2}(E_{1+2})$ versus the energy E_{1+2} according to Eq. (11.2)

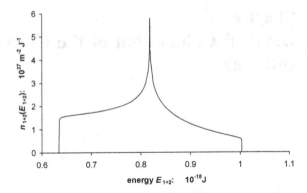

For differentiation of Eq. (11.3) with respect to the strain ε, we have to apply the extended Leibniz rule [1]

$$\frac{d}{d\varepsilon} \int_{x=a(\varepsilon)}^{x=b(\varepsilon)} f(x,\varepsilon)dx = \int_{x=a(\varepsilon)}^{b(\varepsilon)} \frac{\partial f(x,\varepsilon)}{\partial \varepsilon}dx + f(b(\varepsilon),\varepsilon)b'(\varepsilon) - f(a(\varepsilon),\varepsilon)a'(\varepsilon). \quad (11.4)$$

By Herring's equation, we obtain for the surface stress

$$\sigma^{ESB} = \varphi^{ESB} + \frac{\partial \varphi^{ESB}}{\varepsilon_{11}} = \frac{1}{L_1 L_2} \frac{\partial}{\partial \varepsilon_{11}} \int_{E_1^B + E_2^B}^{E_1^T + E_2^T} p(E^S + E_{1+2} - \zeta)(E^S + E_{1+2})n_{1+2}(E_{1+2})dE_{1+2}$$

$$= \frac{1}{L_1 L_2} \left\{ \int_{E_1^B + E_2^B}^{E_1^T + E_2^T} p(E^S + E_{1+2} - \zeta) \left(\frac{\partial E_{1+2}}{\partial \varepsilon_{11}} n_{1+2}(E_{1+2}) + (E^S + E_{1+2}) \frac{\partial n_{1+2}(E_{1+2})}{\partial \varepsilon_{11}} \right) dE_{1+2} \right.$$

$$+ p(E^S + E_1^T + E_2^T - \zeta)(E^S + E_1^T + E_2^T)n_{1+2}(E_1^T + E_2^T) \frac{\partial(E_1^T + E_2^T)}{\partial \varepsilon_{11}}$$

$$\left. - p(E^S + E_1^B + E_2^B - \zeta)(E^S + E_1^B + E_2^B)n_{1+2}(E_1^B + E_2^B) \frac{\partial(E_1^B + E_2^B)}{\partial \varepsilon_{11}} \right\}.$$

$$(11.5)$$

According to Fig. 11.1 it is

$$n_{1+2}\left(E_{1+2}^B\right) = n_{1+2}(E_1^B + E_2^B) = 0 \quad \text{and} \quad n_{1+2}\left(E_{1+2}^T\right) = n_{1+2}\left(E_1^T + E_2^T\right) = 0. \quad (11.6)$$

Fig. 11.2 The area of
integration for the
convolution integral
belonging to the lower energy
band of a two-dimensional
system; The *dashed line* is the
graph for the function
$E_1 + E_2 = E_{1+2}$

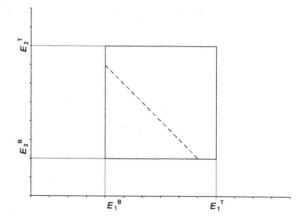

Therefore, we obtain

$$\sigma_{11}^{\mathrm{ESB}} = \varphi^{\mathrm{ESB}} + \frac{\partial \varphi^{\mathrm{ESB}}}{\varepsilon_{11}}$$

$$= \frac{1}{L_1 L_2} \int_{E_1^B + E_2^B}^{E_1^T + E_2^T} p\left(E^S + E_{1+2} - \zeta\right) \left(\frac{\partial E_{1+2}}{\partial \varepsilon_{11}} n_{1+2}(E_{1+2}) + \left(E^S + E_{1+2}\right) \frac{\partial n_{1+2}(E_{1+2})}{\partial \varepsilon_{11}}\right) dE_{1+2} \, .$$

$$(11.7)$$

Figure. 11.2 shows the area of integration for the practical accomplishment of
the convolution procedure according to Eqs. (11.3) and (11.7). For the sake of
simplicity, we suppose

$$E_1^B = E_2^B \quad \text{and} \quad E_1^T = E_2^T \, . \tag{11.8}$$

In Fig. 11.2, the dashed line is the function $E_1 + E_2 = E_{1+2}$. The energy E_{1+2} rises
along the diagonal in the square, from $E_1^B + E_2^B$ to $E_1^T + E_2^T$.

The practical calculations of the surface energy and the surface stress have been
carried out with the following formulae

$$\varphi^{\mathrm{ESB}} = \frac{1}{L_1 L_2} \left\{ \int_{E_1^B + E_2^B}^{E_1^B + E_1^T} p\left(E^S + E_{1+2} - \zeta\right) \int_{E_1^B}^{E_{1+2} - E_1^B} \left(E^S + E_{1+2}\right) n_1(E_1) n_2(E_{1+2} - E_1) dE_1 dE_{1+2} \right.$$

$$+ \int_{E_1^B + E_1^T}^{E_1^T + E_2^T} p\left(E^S + E_{1+2} - \zeta\right) \int_{E_{1+2} - E_1^T}^{E_1^T} \left(E^S + E_{1+2}\right) n_1(E_1) n_2(E_{1+2} - E_1) dE_1 dE_{1+2} \left. \right\} \, .$$

$$(11.9)$$

The application of Herring's formula, Eq. (1.3), to the Eq. (11.9) yields

$$\sigma_{11}^{ESB} = \varphi^{ESB} + \frac{\partial \varphi^{ESB}}{\varepsilon_{11}}$$

$$= \frac{1}{L_1 L_2}\left\{ \frac{\partial}{\partial \varepsilon_{11}} \int_{E_1^B + E_2^B}^{E_1^B + E_1^T} p(E^S + E_{1+2} - \zeta) \int_{E_1^B}^{E_{1+2} - E_1^B} (E^S + E_{1+2}) n_1(E_1) n_2(E_{1+2} - E_1) dE_1 dE_{1+2} \right.$$

$$+ p(E^S + E_1^B + E_1^T - \zeta) \int_{E_1^B}^{E_1^T} (E^S + E_1^B + E_1^T) n_1(E_1) n_2(E_1^B + E_1^T - E_1) dE_1 \frac{\partial(E_1^B + E_1^T)}{\partial \varepsilon_{ii}}$$

$$- p(E^S + E_1^B + E_2^B - \zeta) \int_{E_1^B}^{E_1^B} (E^S + E_1^B + E_2^B) n_1(E_1) n_2(E_1^B + E_2^B - E_1) dE_1 \frac{\partial(E_1^B + E_2^B)}{\partial \varepsilon_{ii}}$$

$$+ \frac{\partial}{\partial \varepsilon_{11}} \int_{E_1^B + E_1^T}^{E_1^T + E_2^T} p(E^S + E_{1+2} - \zeta) \int_{E_{1+2} - E_1^T}^{E_1^T} (E^S + E_{1+2}) n_1(E_1) n_2(E_{1+2} - E_1) dE_1 dE_{1+2}$$

$$+ p(E^S + E_1^T + E_2^T - \zeta) \int_{E_1^T}^{E_1^T} (E^S + E_1^T + E_2^T) n_1(E_1) n_2(E_1^T + E_2^T - E_1) dE_1 \frac{\partial(E_1^T + E_2^T)}{\partial \varepsilon_{ii}}$$

$$\left. - p(E^S + E_1^B + E_1^T - \zeta) \int_{E_1^B}^{E_1^T} (E^S + E_1^B + E_1^T) n_1(E_1) n_2(E_1^B + E_1^T - E_1) dE_1 \frac{\partial(E_1^B + E_1^T)}{\partial \varepsilon_{ii}} \right\}.$$

$$(11.10)$$

As discussed in Chaps. 6 and 8, we differentiate only the eigenvalues E and the densities $n(E,\varepsilon)$ with respect to ε. The third and seventh term counterbalance each other.

$$\sigma_{11}^{ESB} = \frac{1}{L_1 L_2}\left\{ \int_{E_1^B + E_2^B}^{E_1^B + E_1^T} p(E^S + E_{1+2} - \zeta)\left[\int_{E_1^B}^{E_{1+2} - E_1^B} \frac{\partial E_1}{\partial \varepsilon_{11}} n_1(E_1) n_2(E_{1+2} - E_1) \right.\right.$$

$$+ (E^S + E_{1+2})\frac{\partial n_1(E_1)}{\partial \varepsilon_{11}} n_2(E_{1+2} - E_1) dE_1$$

$$+ (E^S + E_{1+2}) n_1(E_{1+2} - E_1^B) n_2(E_1^B)\frac{\partial(E_{1+2} - E_1^B)}{\partial \varepsilon_{11}}$$

$$\left. - (E^S + E_{1+2}) n_1(E_1^B) n_2(E_{1+2} - E_1^B)\frac{\partial(E_1^B)}{\partial \varepsilon_{11}}\right] dE_{1+2}$$

$$+ \int_{E_1^B + E_1^T}^{E_1^T + E_2^T} p(E^S + E_{1+2} - \zeta)\left[\int_{E_{1+2} - E_1^T}^{E_1^T} \frac{\partial E_1}{\partial \varepsilon_{11}} n_1(E_1) n_2(E_{1+2} - E_1) \right.$$

$$+ \left(E^S + E_{1+2}\right) \frac{\partial n_1(E_1)}{\partial \varepsilon_{11}} n_2(E_{1+2} - E_1) dE_1$$

$$+ \left(E^S + E_{1+2}\right) n_1\left(E_1^T\right) n_2\left(E_{1+2} - E_1^T\right) \frac{\partial E_1^T}{\partial \varepsilon_{11}}$$

$$\left. - \left(E^S + E_{1+2}\right) n_1\left(E_{1+2} - E_1^T\right) n_2\left(E_1^T\right) \frac{\partial \left(E_{1+2} - E_1^T\right)}{\partial \varepsilon_{11}} \right] dE_{1+2} \right\}. \tag{11.11}$$

In the case that $n_1(E_1) = n_2(E_1)$, the members in the third and in the last row as well as in the fourth and the seventh row in Eq. (11.11) vanish for the following reasons:

The energy level E^B corresponds to $kc = 0$ but, the smallest value of k is

$$k = \frac{2\pi}{L}. \tag{11.12}$$

Strongly speaking, the energy level E^B is a limit which cannot be reached by the allowed energy levels. Therefore, we have $n(E^B) = 0$.

According to the extended Leibniz rule Eq. (11.4), it follows for the derivation of the total number of energy levels N^{el} with respect to ε

$$\frac{\partial N^{el}}{\partial \varepsilon_{ii}} = \frac{\partial}{\partial \varepsilon_{ii}} \int_{E^B(\varepsilon)}^{E^T(\varepsilon)} n(E_i) dE_i$$

$$= \int_{E^B(\varepsilon)}^{E^T(\varepsilon)} \frac{\partial n(E_i)}{\partial \varepsilon_{ii}} dE_i + n\left(E^T(\varepsilon)\right) \frac{\partial E^T}{\partial \varepsilon_{ii}} - n\left(E^B(\varepsilon)\right) \frac{\partial E^B}{\partial \varepsilon_{ii}}. \tag{11.13}$$

In the infinitely extended body, the number of the energy levels between $E^B(\varepsilon)$ and $E^T(\varepsilon)$ will not be changed due to a dilatation of the solid. That means

$$\frac{\partial}{\partial \varepsilon_{ii}} \int_{E^B(\varepsilon)}^{E^T(\varepsilon)} n(E_i) dE_i = 0. \tag{11.14}$$

Furthermore, also the sum of all changes of n in the energy range between $E^B(\varepsilon)$ and $E^T(\varepsilon)$ is zero, that is

$$\int_{E^B(\varepsilon)}^{E^T(\varepsilon)} \frac{\partial n(E_i)}{\partial \varepsilon_{ii}} dE_i = 0. \tag{11.15}$$

For this reason, it follows

$$n\big(E^T(\varepsilon),\varepsilon\big)\frac{\partial E^T}{\partial \varepsilon_{ii}} = n\big(E^B(\varepsilon),\varepsilon\big)\frac{\partial E^B}{\partial \varepsilon_{ii}}\,. \tag{11.16}$$

We see from Fig. 4.2 and Eq. (4.2) that the quantities $\partial E^T/\partial \varepsilon_{ii}$ and $\partial E^B/\partial \varepsilon_{ii}$ are not zero. Because of $n(E^B(\varepsilon),\varepsilon)=0$ and $\partial E/\partial \varepsilon \neq 0$, $n(E^T(\varepsilon),\varepsilon)=0$ and we obtain

$$
\begin{aligned}
\sigma_{11}^{ESB} = \frac{1}{L_1 L_2}\Bigg\{ &\int\limits_{E_1^B+E_2^B}^{E_1^B+E_1^T} p(E^S+E_{1+2}-\zeta)\left[\int\limits_{E_1^B}^{E_{1+2}-E_1^B}\frac{\partial E_1}{\partial \varepsilon_{11}}n_1(E_1)n_2(E_{1+2}-E_1)dE_1 \right.\\
&+ \int\limits_{E_1^B}^{E_{1+2}-E_1^B}(E^S+E_{1+2})\frac{\partial n_1(E_1)}{\partial \varepsilon_{11}}n_2(E_{1+2}-E_1)dE_1 \Bigg] dE_{1+2}\\
&+ \int\limits_{E_1^B+E_1^T}^{E_1^T+E_2^T} p(E^S+E_{1+2}-\zeta)\left[\int\limits_{E_{1+2}-E_1^T}^{E_1^T}\frac{\partial E_1}{\partial \varepsilon_{11}}n_1(E_1)n_2(E_{1+2}-E_1)dE_1 \right.\\
&+ \int\limits_{E_{1+2}-E_1^T}^{E_1^T}(E^S+E_{1+2})\frac{\partial n_1(E_1)}{\partial \varepsilon_{11}}n_2(E_{1+2}-E_1)dE_1 \Bigg] dE_{1+2} \Bigg\}\,.
\end{aligned}
\tag{11.17}
$$

Also this result corresponds to the Eq. (6.5).

Remark
Haiss [2] considered the derivation of the surface energy with respect to the surface strain at $\varepsilon_{ik}=0$.

References

1. Rothe R (1954) Höhere Mathematik für Mathematiker, Physiker, Ingenieure, Teil II, B. Teubner Verlagsgesellschaft Leipzig, pp S. 143
2. Haiss W (2001) Surface stress on clean and adsorbate-covered solids. Rep Prog Phys 64: 591–648

Chapter 12
Comparison of the Results for the Semi-infinite and the Limited Body

Abstract The results for the semi-infinite and the limited bodies are compared. The findings obtained with the different models for the offspring surface state and the surface free energy are similar but, not identical.

Keyword Surface data from different models

In the foregoing chapters, the allowed energy levels were calculated for a Meander-like potential run in the bulk as it is symbolized in Fig. 3.1. The width of the potential wells a and the width of the potential barriers b in the bulk are $a = b = 2 \times 10^{-10}$ m. The height of the barriers in the bulk amounts to $U_i = 0.8 \cdot 10^{-18}$ J (5 eV) and for the height of the potential step at the surface U^S it has been assumed the value 1.6×10^{-18} J (10 eV).

The precondition for the application of the two models considered in this booklet is the separability of the potential energy of the body. If the system is separable, the chemical potential of the three-dimensional system ζ is the sum of the chemical potentials for the subsystems ζ_i. The Fermi function for the three-dimensional system can be written in the form

$$p(E - \zeta) = \frac{1}{1 + e^{\frac{E_1 + E_2 + E_3 - (\zeta_1 + \zeta_2 + \zeta_3)}{kT}}} \tag{5.19}$$

and the Fermi distribution function for the ith partial system is

$$p(E_i - \zeta_i) = \frac{1}{1 + e^{\frac{E_i - \zeta_i}{kT}}} \tag{5.20}$$

On the condition of Eq. (5.21) even the Eq.

$$p(E - \zeta) = p(E_1 - \zeta_1)p(E_2 - \zeta_2)p(E_3 - \zeta_3) \tag{5.23}$$

holds.

W. Gräfe, *Quantum Mechanical Models of Metal Surfaces and Nanoparticles*,
SpringerBriefs in Applied Sciences and Technology,
DOI 10.1007/978-3-319-19764-7_12

12.1 The Semi-infinite Body

12.1.1 Surface States

In the one-dimensional model for the semi-infinitely extended solids, an offspring surface state appears closely above the lower energy band if the surface energy step U^S exceeds a certain level. For the lattice parameters listed above, this threshold amounts to nearly 1.3×10^{-18} J. In the case mentioned above ($U^S = 1.6 \times 10^{-18}$ J), the energy of the offspring surface state is $E_k^{So} = 0.506 \times 10^{-18}$ J.

For the same parameters, the additional surface state appears at $E_k^{Sa} = 0.678 \times 10^{-18}$ J.

12.1.2 Density Distribution of the Energy Levels

For a one-dimensional system, the density distribution of the energy levels is given in Eq. (2.9) and presented in Fig. 2.4.

The calculation of the density distribution of the energy levels for a two-dimensional system has been carried out by a convolution integral in Chap. 6.

12.1.3 Remark

The calculation of the density distribution for a three-dimensional body should be principally performable by a twofold convolution with a formula of the kind in Eq. (12.1)

$$\int_0^\infty \int_0^\infty n_1(E_1)n_2(E_{1+2} - E_1)dE_1 n_3(E - E_{1+2})dE_{1+2} \qquad (12.1)$$

12.1.4 Surface Free Energy

The contribution to the maximum surface free energy φ^{TR} of a semi-infinite body caused by the electron transitions from the upmost energy level in the bulk E_k^T into the surface state E_k^S amounts to

$$\max\left(\varphi^{TR}\right) = \frac{1}{c^2}\left(E_k^S - E_k^T\right) = \frac{(0.506 \cdot 10^{-18} - 0.502 \cdot 10^{-18})J}{(4 \cdot 10^{-10}m)^2} = 0.025\frac{J}{m^2}$$

$$(12.2)$$

12.2 The Limited Body

Now, we consider one-dimensional models with only ten potential wells.

Originally it was the aim to demonstrate a completely different procedure for an easy and principally correct numerical calculation of the surface parameters in a three-dimensional model.

With the model of a limited solid also the calculation of the wave functions was aspired.

The deduction of the surface free energy proved to be particularly descriptive.

12.2.1 Surface States

With the limited model, it is not possible to demonstrate in a convincing manner the existence of a lower surface state, which is an offspring of the "lower energy band". Instead of the surface state we find only an enhancement of the electron concentration near the surface.

12.2.2 Density Distribution of the Energy Levels

The determination of the distribution of the energy levels has been accomplished by a simple calculation of the triples $E_k + E_i + E_j$ and their sequencing according to their magnitude. This can be realized easily.

The positions of the energy levels for different numbers of the potential wells are presented in Fig. 12.1.

We can see from Fig. 12.1 that for the given values of the potential maxima in the bulk as well as for the potential steps at the surfaces and for a given number N of potential wells only $N - 1$ energy levels fall into the allowed energy band for the bulk of an infinite body. In the model of the limited body, the taking of the remaining energy level as an offspring surface state or as a surface resonance is an arbitrary decision.

However, if each potential well contributes one occupied energy level to the solid, the surface is uncharged if the "lower surface" state is occupied.

Fig. 12.1 The positions of
the energy levels for different
numbers of potential wells

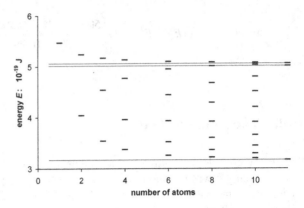

In the one-dimensional model with ten potential wells, the energy eigenvalue of
the additional surface state amounts to 0.8×10^{-18} J.

12.2.3 Surface Free Energy of a Nanocube

For a nanocube with $10 \times 10 \times 10$ potential wells, the dependence of the surface free energy on the chemical potential is depicted in Fig. 8.8. The maximum
value amounts to $0.116 \ \text{J/m}^2$.

12.2.4 Surface Free Energy of a Plate-like Body

The maximum value of surface free energy calculated for a section with
$10 \times 10 \times 10$ potential wells of a plate-like body considered in Chap. 8 also adds up
to $0.116 \ \text{J/m}^2$.

12.3 Summary

The results obtained with the different models for the offspring surface state and the
surface free energy are similar but, not identical.

If in the case of a limited body and empty surface states the chemical potential is
sufficiently decreased, the situation can arise that the surface free energy becomes
negative. This may be deduced from the lowest graph in Fig. 8.4.

The lowest thin line symbolizes the bottom of the lowest allowed energy band
for the model of Kronig and Penney. The next higher line is the top of the allowed
energy band and the highest line gives the position of the offspring surface state.
The rightmost energy levels have been calculated for 100 potential wells.

Chapter 13
Calculation of Surface Stress and Herring's Formula

Abstract The calculation of the surface stress with the Herring's Eq. is analyzed in depth. Without an additional assumption, it is not possible to apply the Herring's equation to the formulae for the surface energy and the surface stress which were deduced in Chap. 8 for the limited body. Let us consider the Eq. (8.2)

$$\varphi^{\text{ESB}} = \frac{1}{2} \frac{1}{L_i L_j} \sum_{\mu'} (E_k^S + E_i^l + E_j^l)_{\mu'} n_{\mu'}^{(2)} \frac{1}{1 + e^{\frac{\left(E_k^S + E_i^l + E_j^l\right)_{\mu'} - \zeta}{kT}}} . \tag{8.2}$$

For the derivation of this formula with respect to the strain ε_{ii}, we have introduced in Chap. 8 an additional restriction for the calculation of the surface stress; it has to be executed the partial derivation of Eq. (8.2). A more detailed analysis of this problem shows that no derivation of the Fermi function with respect to the strain ε_{ii} can appear in the formula for surface stress.

Keyword Applicability of Herring's equation

The formula of Herring, Eq. (1.3), has been deduced by the consideration of an idealized reversible cyclic process. This is described in the paper of Haiss [1]. Therefore, Herring's formula should be of fundamental importance and each result for the surface stress should be in accordance with this formula.

The Herring's Eq. relates the surface free energy to the surface stress, but in Eq. (7.5) we have seen that Herring's Eq. is also applicable to the surface energy caused by the electrons in the surface bands φ^{ESB}.

For the surface energy φ^{ESB}, we write without any doubt

$$\varphi^{\text{ESB}}(\zeta) = \frac{1}{L_i L_j} \int_0^\infty E n^S(E) p(E - \zeta) dE . \tag{13.1}$$

The term $n^S(E)$ means the density of the electron states in the surface bands.

© The Author(s) 2015
W. Gräfe, *Quantum Mechanical Models of Metal Surfaces and Nanoparticles*,
SpringerBriefs in Applied Sciences and Technology,
DOI 10.1007/978-3-319-19764-7_13

By an expansion of the body in the x_i-direction, a surface stress σ_{ii} is generated by the electrons which are located in the near-surface layer. In analogy to the calculations in Chap. 4, we make for this surface stress the ansatz

$$\sigma_{ii} = \frac{F_i^S}{L_j}(\zeta) = \frac{c_i}{L_i L_j} \int\limits_0^\infty f_i(E) n^S(E) p(E - \zeta) dE. \tag{13.2}$$

The computation of this formula has been accomplished without any recourse to the surface energy and to the formula of Herring in Eq. (1.3).

Applying Herring's formula to Eq. (13.1), we have to keep in mind the Leibniz rule and the fact that E is an integration variable and does not depend on strain ε.

In the case the density n^S depends on the strain ε_{ii}, the Eq. (13.2) is not the derivative of Eq. (13.1) and the both Eqs. do not comply with Herring's formula.

Furthermore, without an additional assumption it is not possible to apply the Herring's equation to the formulae for the surface energy and the surface stress which were deduced for the limited body in Chap. 8.

Let us consider the Eq. (8.2)

$$\varphi^{ESB} = \frac{1}{2} \frac{1}{L_i L_j} \sum_{\mu'} (E_k^S + E_i^l + E_j^l)_{\mu'} n_{\mu'}^{(2)} \frac{1}{1 + e^{\frac{(E_k^S + E_i^l + E_j^l)_{\mu'} - \zeta}{kT}}}. \tag{8.2}$$

In Chap. 8, we have introduced an additional restriction for the calculation of the surface stress. It has to be executed the partial derivation of Eq. (8.2) with respect to the strain ε_{ii}. This has been accomplished in a non-stringent manner. By observation of this additional restriction, we have obtained the formula

$$\sigma^{ESB} = \frac{1}{2} \frac{1}{L_i L_j} \left\{ \sum_{\mu''=1}^{M''} \left(\frac{\partial E_i^l}{\partial \varepsilon_{ii}} \right)_{\mu''} \frac{1}{1 + e^{\frac{(E_k^S + E_i^l + E_j^l)_{\mu''} - \zeta}{kT}}} \right\}. \tag{8.7}$$

One would intuitively write Eq. (8.7) because the formula $F = dE/dx$ holds only for potential energies and not for the energies in general. The energy in the Fermi distribution is unspecified.

Let us consider the transition from an infinitely extended body to a limited one. The energy of an infinitely extended chain of atoms is

$$E_i^{1D}(\zeta_i) = \int\limits_{E^B}^{E^T} E_i^e(\eta, b_i) n_i(\eta, b_i) \frac{1}{1 + e^{\frac{\eta - \zeta_i}{kT}}} d\eta. \tag{13.3}$$

The symbol E_i^e means the energy eigenvalues and "1D" marks the one-dimensional body.

In this formula, we substitute the density of the energy levels in an allowed energy band $n(E)$ by a sum of Dirac's delta functions

$$n_i(\eta, b_i) \rightarrow \sum_{v=1}^{N} \delta(\eta - E_{i,v}(b_i)) \tag{13.4}$$

and have

$$E_i^{1D}(\zeta_i) = \int_{-\infty}^{+\infty} \sum_{v=1}^{N} E_{i,v}^e(b_i) \delta(\eta - E_{i,v}(b_i)) \frac{1}{1 + e^{\frac{\eta - \zeta_i}{kT}}} d\eta . \tag{13.5}$$

For an approximation of the δ-function, we use a sequence of normal distributions

$$\delta(\eta - E_{i,v}(b_i), o) = \frac{1}{\sqrt{2\pi}o} e^{-\frac{(\eta - E_{i,v}(b_i))^2}{2o^2}} \tag{13.6}$$

with the parameter o, the Greek letter "omicron," and obtain for the energy

$$E_i^{1D}(\zeta_i) = \lim_{o \to 0} \int_{-\infty}^{+\infty} \sum_{v=1}^{N} E_{i,v}^e(b_i) \frac{1}{\sqrt{2\pi}o} e^{-\frac{(\eta - E_{i,v}(b_i))^2}{2o^2}} \frac{1}{1 + e^{\frac{\eta - \zeta_i}{kT}}} d\eta . \tag{13.7}$$

The quantity o^2 is the variance.

Taking notice of the Leibniz rule, it follows for the approximation of the force in the one-dimensional body

$$F_i^{1D}(\zeta_i) \approx -\frac{d}{db_i} \left\{ \int_{-\infty}^{+\infty} \sum_{v=1}^{N} E_{i,v}^e(b_i) \frac{1}{\sqrt{2\pi}o} e^{-\frac{(\eta - E_{i,v}(b_i))^2}{2o^2}} \frac{1}{1 + e^{\frac{\eta - \zeta_i}{kT}}} d\eta \right\} \tag{13.8}$$

$$
\begin{aligned}
F_i^{1D}(\zeta_i) \approx -\sum_{v=1}^{N} \Bigg\{ &\frac{dE_{i,v}^e(b_i)}{db_i} \int_{-\infty}^{+\infty} \frac{1}{\sqrt{2\pi}o} e^{-\frac{(\eta - E_{i,v}(b_i))^2}{2o^2}} \frac{1}{1 + e^{\frac{\eta - \zeta_i}{kT}}} d\eta \\
&+ E_{i,v}^e(b_i) \int_{-\infty}^{+\infty} \frac{\partial \left(\frac{1}{\sqrt{2\pi}o} e^{-\frac{(\eta - E_{i,v}(b_i))^2}{2o^2}} \right)}{\partial b_i} \frac{1}{1 + e^{\frac{\eta - \zeta_i}{kT}}} d\eta \\
&+ E_{i,v}^e(b_i) \int_{-\infty}^{+\infty} \frac{1}{\sqrt{2\pi}o} e^{-\frac{(\eta - E_{i,v}(b_i))^2}{2o^2}} \frac{\partial}{\partial b_i} \frac{1}{1 + e^{\frac{\eta - \zeta_i}{kT}}} d\eta \Bigg\} .
\end{aligned}
\tag{13.9}
$$

Because η is an independent variable, the last term in Eq. (13.9) vanishes.

In a partial derivation with respect to a given variable, all the other variables have to be kept constant. (For this rule, see Rothe [2] or Smirnow [3]). With x, $y = \text{const}$ we find

$$\frac{\partial f(x,y,z)}{\partial z} \neq \frac{\partial (x(z),y)}{\partial z} \equiv 0. \tag{13.10}$$

Therefore, also the second term in Eq. (13.9) is zero and we obtain for a limited chain of atoms

$$F_i^{1D}(\zeta_i) = -\lim_{o \to 0} \sum_{v=1}^{N} \frac{dE_{i,v}^e(b_i)}{db_i} \int_{-\infty}^{+\infty} \frac{1}{\sqrt{2\pi}o} e^{-\frac{(\eta - E_{i,v}(b_i))^2}{2o^2}} \frac{1}{1 + e^{\frac{\eta - \zeta_i}{kT}}} d\eta. \tag{13.11}$$

Finally we have

$$F_i^{1D}(\zeta_i) = -\sum_{v=1}^{N} \frac{dE_{i,v}^e(b_i)}{db_i} \frac{1}{1 + e^{\frac{E_{i,v} - \zeta_i}{kT}}} \overset{T \to 0}{=} -\sum_{v=1}^{N(\zeta)} \frac{dE_{i,v}^e(b_i)}{db_i}. \tag{13.12}$$

In another calculation method, we start with $E_i^{1D}(T = 0)$ and write

$$F_i^{1D}(\zeta_i) = -\lim_{o \to 0} \frac{d}{db_i} \left\{ E_{i,v}^e(b_i) \int_{-\infty}^{\zeta} \sum_{v=1}^{N} \frac{1}{\sqrt{2\pi}o} e^{-\frac{(\eta - E_{i,v}(b_i))^2}{2o^2}} d\eta \right\}. \tag{13.13}$$

So we find as well

$$F_i^{1D}(\zeta_i) = -\sum_{v=1}^{N(\zeta)} \frac{dE_{i,v}^e(b_i)}{db_i}. \tag{13.14}$$

On the other hand, according to Smirnow [3] it is

$$\frac{\partial}{\partial b_i} \sum_{v=1}^{N(\zeta)} E_{i,v}^e(b_i) p\left(E_{i,v}^e, \zeta\right) = \sum_{v=1}^{N(\zeta)} \left(\frac{\partial E_{i,v}^e(b_i)}{\partial b_i} p\left(E_{i,v}^e, \zeta\right) + E_{i,v}^e \frac{\partial p}{\partial b_i}\right)$$

$$= \sum_{v=1}^{N(\zeta)} \frac{\partial E_{i,v}^e(b_i)}{\partial b_i} p\left(E_{i,v}^e, \zeta\right) \tag{13.15}$$

because the partial derivation $\partial p / \partial b_i$ vanishes.

Consequently, in the Eqs. (8.6) and (8.7) no derivation of the Fermi function with respect to the strain ε_{ii} can appear. (The relation between $\partial \varepsilon_{ii}$ and ∂b_i is given by $\partial \varepsilon_{ii} = \partial b_i / c_i$.)

In contrast to Eq. (13.4), the density of the electron states $n = n(E,b)$ in Eq. (2.9) as well as in Chaps. 5 and 6 depends on the two independent variables E and b!

13.1 Conclusions

If between the energy E and the associated force F the relation

$$F_i = -\frac{dE}{dL_i}.$$ (13.16)

holds, the formula of Shuttleworth, Eq. (1.2), or Herring Eq. (1.3) respectively

$$\sigma_{ii} = \frac{E}{L_i L_j} + \frac{d}{d\varepsilon_{ii}} \frac{E}{L_i L_j} = \frac{E}{L_i L_j} + L_i \frac{d}{dL_i} \frac{E}{L_i L_j} = \frac{1}{L_j} \frac{dE}{dL_i} = -\frac{F_i}{L_j}$$ (13.17)

is fulfilled for the corresponding densities $E/L_i L_j$ and $-F/L_j$.

The force between the atoms in the one-dimensional body correlates with the surface stress on a three-dimensional body according to Eq. (13.18)

$$\sigma_{ii} = -\frac{N_j F_i^{1D}}{L_j} = -\frac{F_i^{1D}}{c_j} = -\frac{F_i}{L_j}.$$ (13.18)

The maximum force exerted by the chain of ten atoms due to a strain of $\varepsilon = 0.5\ \%$ corresponds to the maximum surface stress on a three-dimensional body amounting to $\sigma_{ii} = 2.04$ N/m. See also Fig. 8.10.

It results from the deductions following Eq. (13.3), it is correct if we apply the partial differentiation with respect to ε_{ii} to the Eq. (8.3) in the calculation of the surface stress for a three-dimensional model with a limited number of atoms.

References

1. Haiss W (2001) Surface stress on clean and adsorbate-covered solids. Rep Prog Phys 64: 591–648
2. Rothe R (1954) Höhere Mathematik für Mathematiker, Physiker, Ingenieure, Teil II, B. Teubner Verlagsgesellschaft Leipzig, S. 114
3. Smirnow WI (1956) Lehrgang der höheren Mathematik, Teil I. Deutscher Verlag der Wissenschaften, Berlin, S. 313

Chapter 14
Miscellaneous and Open Questions

Abstract A resume is given and a series of aspects of the surface stress of solids are discussed. The main question is the relation between surface stress and the strength of a solid. According to my theory of fatigue, the defect growth is caused by the migration of particles into the defect due to gradients of tensile stresses. The particle flux can be reduced or stopped by stress gradients of opposite direction. In this way, the near-surface stress should be the cause of the fatigue limit of materials. In accordance with the Eqs. (4.4) and (4.6), the forces acting between the surface atoms should be strong if the corresponding forces in the bulk are strong too. From the existence of a near-surface stress follows a conditional proportionality between the strength of a solid and its surface stress. Another outcome of this book is the fact that an influence of external electric fields on the strength could exist also in such cases in which only electronic processes pass off without any transfer of atoms to or through the interface. The surface free energy reaches its minimum value if the generation of the new surface is an equilibrium process. Therefore, the experimental determination of the surface free energy of brittle materials will be problematic. Due to the generation of a surface on the body the energy levels in the bulk of the solid will also be shifted. A change of the number of electrons in the separated part of a body means the occurrence of a triboelectric process. The number of the potential wells has an influence on the values of the surface parameters.

Keywords Electrocapillarity · Fatigue limit · Surface free energy of solids

14.1 The Scientific Ambition of this Book

It was the main aim of the present book to establish qualitative relationships between the near-surface binding forces and the near-surface local stress on the one hand and the familiar quantities surface charge, surface free energy, and surface stress on the other hand. But it was not the target to determine precisely these last quantities.

© The Author(s) 2015
W. Gräfe, *Quantum Mechanical Models of Metal Surfaces and Nanoparticles*,
SpringerBriefs in Applied Sciences and Technology,
DOI 10.1007/978-3-319-19764-7_14

14.2 Own Results

From a rough estimation of the surface free energy as explained in the text leading to Eq. (7.1) we find for the electron transitions from the lower energy band in the bulk into the lower surface state the surface free energy $\varphi^{TR} \approx 0.7$ J/m^2.

14.2.1 Semi-infinitely Extended Body at 300 K

For the surface energy of the electrons in the surface band φ^{ESB} we have obtained the results presented in Fig. 6.1 ranging from 0 till 8.00 J/m^2.

The data of the surface stress $\sigma_{ii} = \varphi\delta_{ii} + \partial\varphi/\partial\varepsilon_{ii}$ are drawn in Fig. 6.2. They vary from 0 to to 2.11 N/m.

The Fig. 6.3 shows the run of the surface charge density. The values range from 0 till 0.98 C/m^2.

Depending on the position of the Fermi level the surface free energy φ^{TR} resulting from the electron transitions from the bulk into the lower surface band of the lower surface state can add up from 0 to 0.68 J/m^2 as may be seen from Fig. 7.1.

For electron transitions from the lower energy band in bulk to the lower energy band of the upper surface state the surface free energy reaches values up to 6.98 J/m^2 as it may be seen in Fig. 7.2.

The surface stress–charge coefficient (estance) dσ_{ij}/dq is nearly constant and amounts to -1.34 V, see Fig. 9.1.

14.2.2 Nanocube of 10 × 10 × 10 Potential Wells at 0 K

The Fig. 8.7 shows the surface energy of the electrons in the surface band φ^{ESB}. The data rise from 0 to 8.32 J/m^2.

The graph of the surface stress σ^{ESB} is shown in Fig. 8.10. Its maximum value amounts to 1.98 N/m.

The run of the surface charge density q^{ESB} is presented in Fig. 8.12. The values range from 0 to 1.00 C/m^2.

The data of the surface free energy φ calculated for the rise of electron energy levels without regard to the surface states are depicted as a function of the chemical potential in Fig. 8.8. The results rise from 0 to 0.116 J/m^2.

The contribution to the surface free energy φ of the electron transitions from the lower "energy band" in the bulk into the surface state is shown in Fig. 8.9. Here, the values of the energy increase till 10.15 J/m^2.

As shown in Fig. 9.1 the surface stress–charge coefficient (estance) dσ_{ij}/dq varies between -2.13 V and $+0.48$ V in the range of the chemical potential from 1. 1×10^{-18} J till 1.6×10^{-18} J.

Table 14.1 Experimental results

$\varphi(J/m^2)$	σ (N/m)	Estance (V)	Substance	Author
	±0.45		Cu	Vermaak et al. [1–3]
	1.175		Au	
	1.415		Ag	
	2.574		Pt	
	2.1 and 4.2		Ni	Lehwald [4]
		−0.67 till −0.91	Au	Haiss [5]
		−0.83	Au	Ibach [6]
		−0.7 till −1.6	Pt	Weissmüller et al. [7]
		−1 till −1.9	Pt	Viswanath et al. [8]

Table 14.2 Theoretical results

$\varphi(J/m^2)$	σ (N/m)	Estance (V)	Substance	Author
0.49 till 3.25			Bi	Tyson and Miller [9]
			W	
0.26 till 3.65			Na	Miedema [10]
			Rh	
0.62 till 3.036	0.263 till 3.085		Ag	Ackland [11, 12]
			W	
			V	
			Ta	
	1.84 till 2.32		Al	Needs [13]
0.96 till 3.26	0.82 till 5.6		Al, Au, Ir, Pt	Needs et al. [14, 15]
	6.297		Pt	Feibelman [16]
	0.82		Pb	Friesen et al. [17]
	2.77		Au	
	6.297	0 till −1.86	Au	Umeno et al. [18]

If we restrict our considerations to the lower energy band and the lower surface state, the signs and the orders of magnitude of the own surface parameters are in agreement with the experimental and theoretical data obtained by the authors and listed in Tables 14.1 and 14.2.

14.3 Support for the Presented Theory

Viswanath et al. [8] have realized "that the charge accumulated in the space charge region at the metal surface has a pronounced effect on the surface stress of the metal-electrolyte interface." Weissmüller et al. [19] have drawn the conclusion that "by the controlling the net charge in space charge layers at metal surfaces, one can modify the electronic density of states and, thereby, the local properties of the matter at the surface." And the reasoning of Ibach [6] is: "We propose that the surface stress is due to changes in the electronic structure of the substrate surface atoms, even when the surfaces are covered with a high concentration of adsorbed species." These statements support my assumption that the origin of the surface stress and the near-surface stress are exchange forces between the electrons in the energy bands interlinked with the surface states.

14.4 Fatigue Limit

A compressive stress near the surface of a solid can have an influence on strength in different ways. At first, there are well-known technologies to produce compressive stresses in macroscopic ranges of technical goods. As long as tensile stresses caused in the good by external loads do not exceed the compressive stress no fracture will pass off. Such a technical operation can serve for the explanation of the effect of an inherent near-surface stress on the fatigue limit.

On the other hand, according to my theory of fatigue in [20] the defect growth is caused by the migration of particles into the defect due to gradients of tensile stresses. The particle flux can be reduced or stopped by stress gradients of opposite direction. The defects involved in fracture are mainly located at the surface of the bodies. Therefore, an inherent compressive near-surface stress as discussed in relation to Eq. (4.7) should act as a barrier for the defect growth, that is to say, the near-surface stress should be the cause of the fatigue limit of materials not subjected to any technological measure.

A fracture mechanical analysis of the relation between the parameters of the inherent near-surface stress, Eq. (4.7), and the value of the fatigue limit goes beyond the intended scope of this paper. Therefore and for the sake of simplicity, I equate the value of the near-surface local stress immediately at the surface $s_{ii}(0)$ with the magnitude of the fatigue limit.

For one surface state only, the near-surface stress is given by the function

$$s_{ii} = s_{ii}(0)e^{\frac{-2x}{\delta}}. \tag{14.1}$$

For the lower surface state and the lower energy band it results the value of $s_{ii}(0) = 2.44$ GPa. Such a fatigue limit is unrealistically high. But, if we diminish the surface potential step U_S, the attenuation constant δ can rise until infinity and the fatigue limit would intend toward zero. In case the lower surface state has been disappeared, only the higher surface state accounts for the fatigue limit. This surface state depends only weekly on the height of the surface potential step and its attenuation length δ is essentially shorter. In either case, for this surface state a reduction of the fatigue limit can be achieved by a temperature rise, when the electrons leave the lower surface energy band.

14.5 Surface Stress and Young's Modulus

According to Eqs. (4.4) and (4.6), the forces acting between the surface atoms should be strong if the corresponding forces in the bulk are strong too. From the existence of a near-surface stress follows a conditional proportionality between the strength of a solid and its surface stress. However, one has to take into consideration changes in the occupation of the energy levels. Deep in the bulk, the occupation of the energy levels is determined by the temperature and the position of the Fermi level. However, near the surface the energy bands may be bent upwards or downwards. The occupation of the surface states would also depend on this bending. Therefore, there will be no strict proportionality between the binding forces in the bulk on the one hand and the surface stress or the near-surface local stress on the other hand. An additional influence of the surface conditions on the near-surface local stress exists by the attenuation length δ of the surface state wave functions into the bulk.

For some metals the highest values of the theoretical surface stresses calculated with the first principles methods, as compiled by Haiss [21], are depicted versus the Young's modulus Y in Fig. 14.1. The Young's moduli for the isotropic bulk materials are taken from Kohlrausch [22]. The correlation coefficient for the data in Fig. 14.1 amounts to 0.934.

14.6 Electrocapillarity

Another outcome of this book is a qualitative analytical explanation for the fact that an influence of external electric fields on the surface energy (the electrocapillarity), on the surface stress, and on strength could exist also in such cases in which only

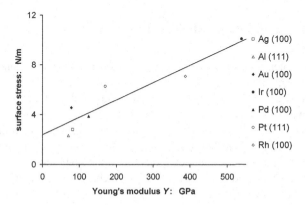

Fig. 14.1 Correlation between the highest values of the surface stresses calculated with the first principles methods for some metals and their Young's moduli Y for the isotropic bulk materials; the correlation coefficient for the data amounts to 0.934.; source: J. Mater. Sci. (2013) 48: 2092–2103

electronic processes pass off without any transfer of atoms to or through the interface.

A consequence of the electrocapillarity for solids is the influence of electric fields on strength or hardness of surfaces. Lichtman et al. [23] have performed measurements of hardness in dependence on the electric potential with the metals thallium, lead, zinc, and tellurium. Pfützenreuter and Masing [24] investigated the rise in the flow rate of lead, zinc, silver, gold, and platinum caused by changes of their polarization in electrolytes. The last authors have interpreted the rise in the flow rate by a reduction in the surface energy of these metals due to the electro-capillarity effect.

14.7 The Minimum of the Surface Free Energy

The surface free energy reaches its minimum value if the generation of the new surface is an equilibrium process. With other words, if the generation of the new surface is a process going on infinitely slowly. The slower this process the higher is the probability that an electron occupying a surface state stems from the highest occupied level of the energy band in the bulk. For high velocities of the surface generation also electrons from deeper energy levels will be transferred into the surface state.

Therefore, the experimental determination of the surface free energy of a brittle solid will be problematic, see Fig. 7.1.

In the generation of a new surface, the transfer probabilities of the electrons will play an important role. Their knowledge will be the key information for a deter-mination of the distribution of the energy levels which are the starting points of the electrons taking part in the surface generation.

14.8 Fermi Level/Chemcal Potential

The chemical potential ζ of a system is determined by the parameters of the res-ervoir which is in contact with the system under consideration.

If we stretch a solid, the energy levels in the solid are shifted and an exchange of electrons between the specimen and the reservoir can go on.

Due to the generation of a surface on the body the energy levels in the solid will be shifted upwards. In order that the number of electrons in a separated part of the body remains unchanged, the chemical potential in the body has to rise as well. (In this case we have to take into consideration a change from a macrocanonical ensemble to a canonical one.) A change of the number of electrons in the separated part of a body means the occurrence of a triboelectric process.

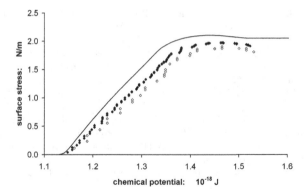

Fig. 14.2 The dependence of the surface stress on the chemical potential calculated for a limited body with 10 × 10 × 10 potential wells (*full rhombs*) and for a cube with 6 × 6 × 6 atoms (*empty symbols*), both at $T = 0$ K, as well as for an infinitely extended body (*line*) at $T = 300$ K

14.9 The Influence of the Number of Atoms on the Results

The number of the potential wells has an influence on the values of the surface parameters. All the values of the surface stress diminish by a reduction of the number of atoms. In the considered models the cause of the forces are the alterations of the height of the potential wells in the bulk. With the reduction of the number of atoms the influence of the surface potential steps on the energy eigenvalues becomes more predominant and, paradoxically, the effect of the strain on the surface stress decreases. The Fig. 14.2 shows once more the data depicted in Fig. 8.10 for the surface stress of an infinitely extended body at $T = 300$ K and for a cube with 10 × 10 × 10 atoms at $T = 0$ K. Furthermore, we see in Fig. 14.2 the corresponding results for a nanocube with only 6 × 6 × 6 potential wells.

References

1. Mays CW, Vermaak JS, Kuhlmann-Wilsdorf D (1968) On surface stress and surface tension: II. Determination of the surface stress of gold. Surf Sci 12:134–140
2. Wassermann HJ, Vermaak JS (1970) On the determination of lattice contraction in very small silver particles. Surf Sci 22:164–172
3. Wassermann HJ, Vermaak JS (1972) On the determination of the surface stress of copper and platinium. Surf Sci 32:168–174
4. Lehwald S, Wolf F, Ibach H, Hall BM, Mills DL (1987) Surface vibrations on Ni(110): the role of surface stress. Surf Sci 192:131–162
5. Haiss W, Nichols RJ, Sass JK, Charle KP (1998) Linear correlation between surface stress and surface charge in anion adsorption on Au(111). J Electroanal Chem 452:199–202
6. Ibach H (1999) Stress in densely packed adsorbate layers and stress at the solid-liquid interface – Is the stress due to repulsive interactions between the adsorbed species? Electrochim Acta 45:575–581

7. Weissmüller J, Viswanath RN, Kramer D, Zimmer P, Würschum R, Gleiter H (2003) Charge-induced reversible strain. Science 300:312–315
8. Viswanath RN, Kramer D, Weissmüller J (2005) Variation of the surface stress-charge coefficient of platinum with electrolyte concentration. Langmuir 21:4604–4609
9. Tyson WR, Miller WA (1977) Surface free energies of solid metals: estimation from liquid surface tension. Surf Sci 62:267–276
10. Miedema AR (1979) Das Atom als Baustein in der Metallkunde. Philips Techn Rundsch 38:269–281
11. Ackland GJ, Finnis MW (1986) Semi-empirical calculation of solid surface tensions in body-centred cubic transition metals. Phil Mag A 54:301–315
12. Ackland GJ, Tichy G, Vitek V, Finnis MW (1987) Simple N-body potentials for the noble metals and nickel. Phil Mag A 56:735–756
13. Needs RJ (1987) Calculations of the surface stress tensor at aluminum (111) and (110) surfaces. Phys Rev Lett 58:53–56
14. Needs RJ, Godfrey MJ (1990) Surface stress of aluminum and jellium. Phys Rev B 42: 10933–10939
15. Needs RJ, Godfrey MJ, Mansfield M (1991) Theory of surface stress and surface reconstruction. Surf Sci 242:215–221
16. Feibelman PJ (1997-II) First-principles calculations of stress induced by gas adsorption on Pt (111). Phys Rev B 56:2175–2182
17. Friesen C, Dimitrov N, Cammarata RC, Siradzki K (2001) Surface stress and electrocapillarity of solid electrodes. Langmuir 17:807–815
18. Umeno Y, Elsässer C, Meyer B, Gumbsch P, Nothacker M, Weissmüller J, Evers F (2007) Ab initio study of surface stress response to charging, EPL 78 13001-p1–13001-p-5
19. Kramer D, Viswanath RN, Weissmüller J (2004) Surface-stress induced macroscopic bending of nanoporous gold cantilevers. Nano Lett 4:793–796
20. Gräfe W (1989) A surface-near stress resulting from Tamm's surface states. Cryst Res Technol 24:879–886
21. Haiss W (2001) Surface stress on clean and adsorbate-covered solids. Rep Prog Phys 64: 591–648
22. Hahn D, Wagner S (eds) (1986) F. Kohlrausch, Praktische Physik, Bd.3, Teubner, Stuttgart, p 40
23. Lichtman WI, Rehbinder PA, Karpenko GW (1964) Der Einfluss Grenzflächenaktiver Stoffe auf die Deformation von Metallen, p 54. Akademie-Verlag, Berlin
24. Pfützenreuter A, Masing G (1951) Zunahme der Geschwindigkeit des plastischen Fließens von Metallen im Elektrolyten bei der elektrochemischen Polarisation. Zeitschr. f. Metallk. 42:361–370

Index

© The Author(s) 2015
W. Gräfe, *Quantum Mechanical Models of Metal Surfaces and Nanoparticles*,
SpringerBriefs in Applied Sciences and Technology,
DOI 10.1007/978-3-319-19764-7

Printed in the United States
By Bookmasters